Honda NE/NB50 Vision & SA50 Vision Met-in Owners Workshop Manual

by Pete Shoemark
with an additional Chapter on the SA50 Vision Met-in
by Penny Cox

Models covered
NE50M Vision. 49cc. April 1985 to March 1990
NB50M Vision-X. 49cc. April 1985 to August 1987
NE50TH Vision. 49cc. August 1987 to January 1988
SA50 Vision Met-in. 49cc. October 1988 to November 1995
Note: This Manual does not cover the NT50 Mini Vision models

(1278-1X3)

THE BOOK

Haynes Group Limited
Haynes North America, Inc

www.haynes.com

Acknowledgements

Our thanks are due to Paul Branson Motorcycles of Yeovil, Somerset, who provided the NE50 Vision and SA50 Vision Met-in featured in this Manual. The Avon Rubber Company supplied information on tyre care and fitting, and NGK Spark Plugs (UK) Ltd provided information on plug maintenance and electrode conditions.

A book in the **Haynes Owners Workshop Manual Series**

ISBN 978 1 85960 497 7

British Library Cataloguing in Publication Data

A catalogue record for this book is available from the British Library

Contents

Page

Introductory pages
About this manual   5
Introduction to the Honda NE and NB50 Vision   5
Model dimensions and weights   5
Ordering spare parts   6
Safety first!   7
Tools and working facilities   8
Fault diagnosis   11
Routine maintenance   18

Chapter 1
Engine and transmission   25

Chapter 2
Fuel system and lubrication   59

Chapter 3
Ignition system   68

Chapter 4
Frame and suspension   75

Chapter 5
Wheels, brakes and tyres   90

Chapter 6
Electrical system   99

Chapter 7
The SA50 Vision Met-in model   114

Wiring diagrams   124

Conversion factors   126

Index   127

The Honda NE50 Vision

Engine/transmission unit

About this manual

The purpose of this manual is to present the owner with a concise and graphic guide which will enable him to tackle any operation from basic routine maintenance to a major overhaul. It has been assumed that any work would be undertaken without the luxury of a well-equipped workshop and a range of manufacturer's service tools.

To this end, the machine featured in the manual was stripped and rebuilt in our own workshop, by a team comprising a mechanic, a photographer and the author. The resulting photographic sequence depicts events as they took place, the hands shown being those of the author and the mechanic.

The use of specialised, and expensive, service tools was avoided unless their use was considered to be essential due to risk of breakage or injury. There is usually some way of improvising a method of removing a stubborn component, providing that a suitable degree of care is exercised.

The author learnt his motorcycle mechanics over a number of years, faced with the same difficulties and using similar facilities to those encountered by most owners. It is hoped that this practical experience can be passed on through the pages of this manual.

Where possible, a well-used example of the machine is chosen for the workshop project, as this highlights any areas which might be particularly prone to giving rise to problems. In this way, any such difficulties are encountered and resolved before the text is written, and the techniques used to deal with them can be incorporated in the relevant section. Armed with a working knowledge of the machine, the author undertakes a considerable amount of research in order that the maximum amount of data can be included in the manual.

A comprehensive section, preceding the main part of the manual, describes procedures for carrying out the routine maintenance of the machine at intervals of time and mileage. This section is included particularly for those owners who wish to ensure the efficient day-to-day running of their motorcycle, but who choose not to undertake overhaul or renovation work.

Each Chapter is divided into numbered sections. Within these sections are numbered paragraphs. Cross reference throughout the manual is quite straightforward and logical. When reference is made 'See Section 6.10' it means Section 6, paragraph 10 in the same Chapter. If another Chapter were intended, the reference would read, for example, 'See Chapter 2, Section 6.10'. All the photographs are captioned with a section/paragraph number to which they refer and are relevant to the Chapter text adjacent.

Figures (usually line illustrations) appear in a logical but numerical order, within a given Chapter. Fig. 1.1 therefore refers to the first figure in Chapter 1.

Left-hand and right-hand descriptions of the machines and their components refer to the left and right of a given machine when the rider is seated normally.

Motorcycle manufacturers continually make changes to specifications and recommendations, and these, when notified, are incorporated into our manuals at the earliest opportunity.

Introduction to the Honda NE and NB50 Vision

The Honda Vision models are the latest in a line of scooter-styled mopeds offered by the various Japanese factories and reflect the changing tastes of this sector of the motorcycle market. In the UK in particular, a change in the definition of a moped has meant that the power and speed of all such machines is restricted, and the requirement for pedals to be fitted has been dropped. As a result there is little to commend a moped designed along conventional motorcycle lines, and indeed fitting four- or five-speed gearboxes becomes a pointless exercise when one or two gears works better.

The Vision models provide something much closer to the original concept of a moped; a simple powered alternative to the bicycle as a means of economical transport. The overall layout is derived from the Italian scooter designs, with the engine and transmission hidden behind body panels. The step-through frame and low seat height allow the machine to be used by male or female riders alike, and the front legshield provides some measure of weather protection for the rider. Both models have a lockable storage compartment inside the legshield and a small luggage rack at the rear, whilst the Vision-X (NB50 M) model has additional lockable side compartments. A range of optional luggage and weather protection equipment is available.

In addition to the original Vision and Vision-X models introduced in 1985, a limited edition model was released in 1987, named the NE50TH. The TH model is virtually identical to the standard NE50 Vision except for its seat covering (which has a check pattern on the rear section), its black body panels with a crest on the rear panels, and gold pinstriping.

Mechanically, the Vision models are far more sophisticated than the simple controls might suggest. A great deal of effort has been directed at making the machines as easy to use as possible, and this in turn means that areas like manual starting, clutch, gearchange and choke are replaced by fully automatic alternatives. From the rider's point of view the machine could hardly be simpler to ride. All controls are confined to the handlebar area and consist of two bicycle-type brake levers, a throttle control and the electrical switches.

The NE50 Vision, NB50 Vision-X and NE50TH Vision are all covered in the main text of the manual. For information relating to the SA50 Vision Met-in, refer to Chapter 7.

Model dimensions and weights

Refer to Chapter 7 for information relating to the SA50 Met-in

	NE 50 Vision	NB 50 Vision-X
Overall length	1585 mm (62.4 in)	1595 mm (62.8 in)
Overall width	625 mm (24.6 in)	625 mm (24.6 in)
Overall height	965 mm (38.0 in)	965 mm (38.0 in)
Wheelbase	1130 mm (44.5 in)	1130 mm (44.5 in)
Ground clearance	100 mm (3.9 in)	100 mm (3.9 in)
Seat height	690 mm (27.2 in)	690 mm (27.2 in)
Dry weight	53.1 kg (117.1 lb)	54.4 kg (119.9 lb)

Ordering spare parts

When replacement parts are required for your Honda, it is advisable to deal with an official Honda dealer. He is in the best position to offer specialist advice and will be able to supply the more commonly used parts from stock. If the parts need to be ordered, remember that an official dealer will be able to arrange faster delivery than a non-specialist supplier. Try to order parts well in advance where this is possible. For example, read through the appropriate section of the manual and see whether gaskets or seals will be needed. This can often avoid having the machine off the road for a week or two while they are ordered.

When ordering, always quote the machine details in full. This will ensure that the correct parts are supplied and will take into account any retrospective manufacturer's modifications. You will need the frame number, which is stamped into a metal plate on the frame, below the left-hand edge of the footboard, and the engine number, which is stamped into the transmission casing to the left of the rear wheel.

During the initial warranty period, and as a general rule, make sure that only genuine Honda parts are used. Fitting non-standard parts may well invalidate the warranty, and more importantly, could prove dangerous. Be particularly wary of pattern safety-related parts such as brake and suspension components. These often resemble the original parts very closely and may even be supplied in counterfeit packaging and sold as genuine items.

Some of the more consumable items, such as spark plugs, bulbs, oils, greases and tyres can be purchased from local sources like accessory shops and motor factors, or from mail order suppliers. Always stick to well-known and reputable brands and make sure that the items supplied are suitable to your machine. When buying tyres, be warned that some perfectly good makes of tyre may not be suited to the suspension characteristics of your model. A Honda dealer or any good tyre supplier will be able to advise here; if they seem vague or non-commital, go elsewhere.

Frame number is stamped on plate below left-hand edge of the footboard

Engine number is stamped into rear of transmission casing

Safety first!

Professional motor mechanics are trained in safe working procedures. However enthusiastic you may be about getting on with the job in hand, do take the time to ensure that your safety is not put at risk. A moment's lack of attention can result in an accident, as can failure to observe certain elementary precautions.

There will always be new ways of having accidents, and the following points do not pretend to be a comprehensive list of all dangers; they are intended rather to make you aware of the risks and to encourage a safety-conscious approach to all work you carry out on your vehicle.

Essential DOs and DON'Ts

DON'T start the engine without first ascertaining that the transmission is in neutral.

DON'T suddenly remove the filler cap from a hot cooling system – cover it with a cloth and release the pressure gradually first, or you may get scalded by escaping coolant.

DON'T attempt to drain oil until you are sure it has cooled sufficiently to avoid scalding you.

DON'T grasp any part of the engine, exhaust or silencer without first ascertaining that it is sufficiently cool to avoid burning you.

DON'T allow brake fluid or antifreeze to contact the machine's paintwork or plastic components.

DON'T syphon toxic liquids such as fuel, brake fluid or antifreeze by mouth, or allow them to remain on your skin.

DON'T inhale dust – it may be injurious to health (see *Asbestos* heading).

DON'T allow any spilt oil or grease to remain on the floor – wipe it up straight away, before someone slips on it.

DON'T use ill-fitting spanners or other tools which may slip and cause injury.

DON'T attempt to lift a heavy component which may be beyond your capability – get assistance.

DON'T rush to finish a job, or take unverified short cuts.

DON'T allow children or animals in or around an unattended vehicle.

DON'T inflate a tyre to a pressure above the recommended maximum. Apart from overstressing the carcase and wheel rim, in extreme cases the tyre may blow off forcibly.

DO ensure that the machine is supported securely at all times. This is especially important when the machine is blocked up to aid wheel or fork removal.

DO take care when attempting to slacken a stubborn nut or bolt. It is generally better to pull on a spanner, rather than push, so that if slippage occurs you fall away from the machine rather than on to it.

DO wear eye protection when using power tools such as drill, sander, bench grinder etc.

DO use a barrier cream on your hands prior to undertaking dirty jobs – it will protect your skin from infection as well as making the dirt easier to remove afterwards; but make sure your hands aren't left slippery. Note that long-term contact with used engine oil can be a health hazard.

DO keep loose clothing (cuffs, tie etc) and long hair well out of the way of moving mechanical parts.

DO remove rings, wristwatch etc, before working on the vehicle – especially the electrical system.

DO keep your work area tidy – it is only too easy to fall over articles left lying around.

DO exercise caution when compressing springs for removal or installation. Ensure that the tension is applied and released in a controlled manner, using suitable tools which preclude the possibility of the spring escaping violently.

DO ensure that any lifting tackle used has a safe working load rating adequate for the job.

DO get someone to check periodically that all is well, when working alone on the vehicle.

DO carry out work in a logical sequence and check that everything is correctly assembled and tightened afterwards.

DO remember that your vehicle's safety affects that of yourself and others. If in doubt on any point, get specialist advice.

IF, in spite of following these precautions, you are unfortunate enough to injure yourself, seek medical attention as soon as possible.

Asbestos

Certain friction, insulating, sealing, and other products – such as brake linings, clutch linings, gaskets, etc – contain asbestos. *Extreme care must be taken to avoid inhalation of dust from such products since it is hazardous to health.* If in doubt, assume that they *do* contain asbestos.

Fire

Remember at all times that petrol (gasoline) is highly flammable. Never smoke, or have any kind of naked flame around, when working on the vehicle. But the risk does not end there – a spark caused by an electrical short-circuit, by two metal surfaces contacting each other, by careless use of tools, or even by static electricity built up in your body under certain conditions, can ignite petrol vapour, which in a confined space is highly explosive.

Always disconnect the battery earth (ground) terminal before working on any part of the fuel or electrical system, and never risk spilling fuel on to a hot engine or exhaust.

It is recommended that a fire extinguisher of a type suitable for fuel and electrical fires is kept handy in the garage or workplace at all times. Never try to extinguish a fuel or electrical fire with water.

Note: *Any reference to a 'torch' appearing in this manual should always be taken to mean a hand-held battery-operated electric lamp or flashlight. It does not mean a welding/gas torch or blowlamp.*

Fumes

Certain fumes are highly toxic and can quickly cause unconsciousness and even death if inhaled to any extent. Petrol (gasoline) vapour comes into this category, as do the vapours from certain solvents such as trichloroethylene. Any draining or pouring of such volatile fluids should be done in a well ventilated area.

When using cleaning fluids and solvents, read the instructions carefully. Never use materials from unmarked containers – they may give off poisonous vapours.

Never run the engine of a motor vehicle in an enclosed space such as a garage. Exhaust fumes contain carbon monoxide which is extremely poisonous; If you need to run the engine, always do so in the open air or at least have the rear of the vehicle outside the workplace.

The battery

Never cause a spark, or allow a naked light, near the vehicle's battery. It will normally be giving off a certain amount of hydrogen gas, which is highly explosive.

Always disconnect the battery earth (ground) terminal before working on the fuel or electrical systems.

If possible, loosen the filler plugs or cover when charging the battery from an external source. Do not charge at an excessive rate or the battery may burst.

Take care when topping up and when carrying the battery. The acid electrolyte, even when diluted, is very corrosive and should not be allowed to contact the eyes or skin.

If you ever need to prepare electrolyte yourself, always add the acid slowly to the water, and never the other way round. Protect against splashes by wearing rubber gloves and goggles.

Mains electricity and electrical equipment

When using an electric power tool, inspection light etc, always ensure that the appliance is correctly connected to its plug and that, where necessary, it is properly earthed (grounded). Do not use such appliances in damp conditions and, again, beware of creating a spark or applying excessive heat in the vicinity of fuel or fuel vapour. Also ensure that the appliances meet the relevant national safety standards.

Ignition HT voltage

A severe electric shock can result from touching certain parts of the ignition system, such as the HT leads, when the engine is running or being cranked, particularly if components are damp or the insulation is defective. Where an electronic ignition system is fitted, the HT voltage is much higher and could prove fatal.

Tools and working facilities

The first priority when undertaking maintenance or repair work of any sort on a motorcycle is to have a clean, dry, well-lit working area. Work carried out in peace and quiet in the well-ordered atmosphere of a good workshop will give more satisfaction and much better results than can usually be achieved in poor working conditions. A good workshop must have a clean flat workbench or a solidly constructed table of convenient working height. The workbench or table should be equipped with a vice which has a jaw opening of at least 4 in (100 mm). A set of jaw covers should be made from soft metal such as aluminium alloy or copper, or from wood. These covers will minimise the marking or damaging of soft or delicate components which may be clamped in the vice. Some clean, dry, storage space will be required for tools, lubricants and dismantled components. It will be necessary during a major overhaul to lay out engine/gearbox components for examination and to keep them where they will remain undisturbed for as long as is necessary. To this end it is recommended that a supply of metal or plastic containers of suitable size is collected. A supply of clean, lint-free, rags for cleaning purposes and some newspapers, other rags, or paper towels for mopping up spillages should also be kept. If working on a hard concrete floor note that both the floor and one's knees can be protected from oil spillages and wear by cutting open a large cardboard box and spreading it flat on the floor under the machine or workbench. This also helps to provide some warmth in winter and to prevent the loss of nuts, washers, and other tiny components which have a tendency to disappear when dropped on anything other than a perfectly clean, flat, surface.

Unfortunately, such working conditions are not always available to the home mechanic. When working in poor conditions it is essential to take extra time and care to ensure that the components being worked on are kept scrupulously clean and to ensure that no components or tools are lost or damaged.

A selection of good tools is a fundamental requirement for anyone contemplating the maintenance and repair of a motor vehicle. For the owner who does not possess any, their purchase will prove a considerable expense, offsetting some of the savings made by doing-it-yourself. However, provided that the tools purchased meet the relevant national safety standards and are of good quality, they will last for many years and prove an extremely worthwhile investment.

To help the average owner to decide which tools are needed to carry out the various tasks detailed in this manual, we have compiled three lists of tools under the following headings: *Maintenance and minor repair, Repair and overhaul,* and *Specialized.* The newcomer to practical mechanics should start off with the simpler jobs around the vehicle. Then, as his confidence and experience grow, he can undertake more difficult tasks, buying extra tools as and when they are needed. In this way, a *Maintenance and minor repair* tool kit can be built-up into a *Repair and overhaul* tool kit over a considerable period of time without any major cash outlays. The experienced home mechanic will have a tool kit good enough for most repair and overhaul procedures and will add tools from the specialized category when he feels the expense is justified by the amount of use these tools will be put to.

It is obviously not possible to cover the subject of tools fully here. For those who wish to learn more about tools and their use there is a book entitled *Motorcycle Workshop Practice Manual* (Bk No 1454).

As a general rule, it is better to buy the more expensive, good quality tools. Given reasonable use, such tools will last for a very long time, whereas the cheaper, poor quality, item will wear out faster and need to be renewed more often, thus nullifying the original saving. There is also the risk of a poor quality tool breaking while in use, causing personal injury or expensive damage to the component being worked on.

For practically all tools, a tool factor is the best source since he will have a very comprehensive range compared with the average garage or accessory shop. Having said that, accessory shops often offer excellent quality tools at discount prices, so it pays to shop around. There are plenty of tools around at reasonable prices, but always aim to purchase items which meet the relevant national safety standards. If in doubt, seek the advice of the shop proprietor or manager before making a purchase.

The basis of any toolkit is a set of spanners. While open-ended spanners with their slim jaws, are useful for working on awkwardly-positioned nuts, ring spanners have advantages in that they grip the nut far more positively. There is less risk of the spanner slipping off the nut and damaging it, for this reason alone ring spanners are to be preferred. Ideally, the home mechanic should acquire a set of each, but if expense rules this out a set of combination spanners (open-ended at one end and with a ring of the same size at the other) will provide a good compromise. Another item which is so useful it should be considered an essential requirement for any home mechanic is a set of socket spanners. These are available in a variety of drive sizes. It is recommended that the ½-inch drive type is purchased to begin with as although bulkier and more expensive than the ⅜-inch type, the larger size is far more common and will accept a greater variety of torque wrenches, extension pieces and socket sizes. The socket set should comprise sockets of sizes between 8 and 24 mm, a reversible ratchet drive, an extension bar of about 10 inches in length, a spark plug socket with a rubber insert, and a universal joint. Other attachments can be added to the set at a later date.

Maintenance and minor repair tool kit

Set of spanners 8 – 24 mm
Set of sockets and attachments
Spark plug spanner with rubber insert – 10, 12, or 14 mm as appropriate
Adjustable spanner
C-spanner/pin spanner
Torque wrench (same size drive as sockets)
Set of screwdrivers (flat blade)
Set of screwdrivers (cross-head)
Set of Allen keys 4 – 10 mm
Impact screwdriver and bits
Ball pein hammer – 2 lb
Hacksaw (junior)
Self-locking pliers – Mole grips or vice grips
Pliers – combination
Pliers – needle nose
Wire brush (small)
Soft-bristled brush
Tyre pump
Tyre pressure gauge
Tyre tread depth gauge
Oil can
Fine emery cloth
Funnel (medium size)
Drip tray
Grease gun
Set of feeler gauges
Strobe timing light
Continuity tester (dry battery and bulb)
Soldering iron and solder
Wire stripper or craft knife
PVC insulating tape
Assortment of split pins, nuts, bolts, and washers

Repair and overhaul toolkit

The tools in this list are virtually essential for anyone undertaking major repairs to a motorcycle and are additional to the tools listed above. Concerning Torx driver bits, Torx screws are encountered on some of the more modern machines where their use is restricted to fastening certain components inside the engine/gearbox unit. It is therefore recommended that if Torx bits cannot be borrowed from a local dealer, they are purchased individually as the need arises. They are not in regular use in the motor trade and will therefore only be available in specialist tool shops.

Plastic or rubber soft-faced mallet
Torx driver bits
Pliers – electrician's side cutters
Circlip pliers – internal (straight or right-angled tips are available)
Circlip pliers – external
Cold chisel
Centre punch
Pin punch
Scriber
Scraper (made from soft metal such as aluminium or copper)
Soft metal drift
Steel rule/straight edge
Assortment of files

Electric drill and bits
Wire brush (large)
Soft wire brush (similar to those used for cleaning suede shoes)
Sheet of plate glass
Hacksaw (large)
Stud extractor set (E-Z out)

Specialized tools

This is not a list of the tools made by the machine's manufacturer to carry out a specific task on a limited range of models. Occasional references are made to such tools in the text of this manual and, in general, an alternative method of carrying out the task without the manufacturer's tool is given where possible. The tools mentioned in this list are those which are not used regularly and are expensive to buy in view of their infrequent use. Where this is the case it may be possible to hire or borrow the tools against a deposit from a local dealer or tool hire shop. An alternative is for a group of friends or a motorcycle club to join in the purchase.

Piston ring compressor
Universal bearing puller
Cylinder bore honing attachment (for electric drill)
Micrometer set
Vernier calipers
Dial gauge set
Cylinder compression gauge
Multimeter
Dwell meter/tachometer

Care and maintenance of tools

Whatever the quality of the tools purchased, they will last much longer if cared for. This means in practice ensuring that a tool is used for its intended purpose; for example screwdrivers should not be used as a substitute for a centre punch, or as chisels. Always remove dirt or grease and any metal particles but remember that a light film of oil will prevent rusting if the tools are infrequently used. The common tools can be kept together in a large box or tray but the more delicate, and more expensive, items should be stored separately where they cannot be damaged. When a tool is damaged or worn out, be sure to renew it immediately. It is false economy to continue to use a worn spanner or screwdriver which may slip and cause expensive damage to the component being worked on.

Fastening systems

Fasteners, basically, are nuts, bolts and screws used to hold two or more parts together. There are a few things to keep in mind when working with fasteners. Almost all of them use a locking device of some type; either a lock washer, lock nut, locking tab or thread adhesive. All threaded fasteners should be clean, straight, have undamaged threads and undamaged corners on the hexagon head where the spanner fits. Develop the habit of replacing all damaged nuts and bolts with new ones.

Rusted nuts and bolts should be treated with a rust penetrating fluid to ease removal and prevent breakage. After applying the rust penetrant, let it 'work' for a few minutes before trying to loosen the nut or bolt. Badly rusted fasteners may have to be chiseled off or removed with a special nut breaker, available at tool shops.

Flat washers and lock washers, when removed from an assembly, should always be replaced exactly as removed. Replace any damaged washers with new ones. Always use a flat washer between a lock washer and any soft metal surface (such as aluminium), thin sheet metal or plastic. Special lock nuts can only be used once or twice before they lose their locking ability and must be renewed.

If a bolt or stud breaks off in an assembly, it can be drilled out and removed with a special tool called an E-Z out. Most dealer service departments and motorcycle repair shops can perform this task, as well as others (such as the repair of threaded holes that have been stripped out).

Spanner size comparison

Jaw gap (in)	Spanner size	Jaw gap (in)	Spanner size
0.250	1/4 in AF	0.945	24 mm
0.276	7 mm	1.000	1 in AF
0.313	5/16 in AF	1.010	9/16 in Whitworth; 5/8 in BSF
0.315	8 mm	1.024	26 mm
0.344	11/32 in AF; 1/8 in Whitworth	1.063	11/16 in AF; 27 mm
0.354	9 mm	1.100	5/16 in Whitworth; 11/16 in BSF
0.375	3/8 in AF	1.125	11/8 in AF
0.394	10 mm	1.181	30 mm
0.433	11 mm	1.200	11/16 in Whitworth; 3/4 in BSF
0.438	7/16 in AF	1.250	11/4 in AF
0.445	3/16 in Whitworth; 1/4	1.260	32 mm
0.472	12 mm	1.300	3/4 in Whitworth; 7/8 in BSF
0.500	1/2 in AF	1.313	15/16 in AF
0.512	13 mm	1.390	13/16 in Whitworth; 15/16 in BSF
0.525	1/4 in Whitworth; 5/16 in BSF	1.417	36 mm
0.551	14 mm	1.438	17/16 in AF
0.563	9/16 in AF	1.480	7/8 in Whitworth; 1 in BSF
0.591	15 mm	1.500	11/2 in AF
0.600	5/16 in Whitworth; 3/8 in BSF	1.575	40 mm; 15/16 in Whitworth
0.625	5/8 in AF	1.614	41 mm
0.630	16 mm	1.625	15/8 in AF
0.669	17 mm	1.670	1 in Whitworth; 11/8 in BSF
0.686	11/16 in AF	1.688	111/16 in AF
0.709	18 mm	1.811	46 mm
0.710	3/8 in Whitworth; 7/16 in BSF	1.813	113/16 in AF
0.748	19 mm	1.860	11/8 in Whitworth; 11/4 in BSF
0.750	3/4 in AF	1.875	17/8 in AF
0.813	13/16 in AF	1.969	50 mm
0.820	7/16 in Whitworth; 1/2 in BSF	2.000	2 in AF
0.866	22 mm	2.050	11/4 in Whitworth; 13/8 in BSF
0.875	7/8 in AF	2.165	55 mm
0.920	1/2 in Whitworth; 9/16 in BSF	2.362	60 mm
0.938	15/16 in AF		

Standard torque settings

Specific torque settings will be found at the end of the specifications section of each chapter. Where no figure is given, bolts should be secured according to the table below.

Fastener type (thread diameter)	kgf m	lbf ft
5mm bolt or nut	0.45 – 0.6	3.5 – 4.5
6 mm bolt or nut	0.8 – 1.2	6 – 9
8 mm bolt or nut	1.8 – 2.5	13 – 18
10 mm bolt or nut	3.0 – 4.0	22 – 29
12 mm bolt or nut	5.0 – 6.0	36 – 43
5 mm screw	0.35 – 0.5	2.5 – 3.6
6 mm screw	0.7 – 1.1	5 – 8
6 mm flange bolt	1.0 – 1.4	7 – 10
8 mm flange bolt	2.4 – 3.0	17 – 22
10 mm flange bolt	3.5 – 4.5	25 – 33

Fault diagnosis

Contents

Introduction 1

Engine does not start when turned over
No fuel flow to carburettor 2
Fuel not reaching cylinder 3
Engine flooding 4
No spark at plug 5
Weak spark at plug 6
Compression low 7

Engine stalls after starting
General causes 8

Poor running at idle and low speed
Weak spark at plug or erratic firing 9
Fuel/air mixture incorrect 10
Compression low 11

Acceleration poor
General causes 12

Poor running or lack of power at high speeds
Weak spark at plug or erratic firing 13
Fuel/air mixture incorrect 14
Compression low 15

Knocking or pinking
General causes 16

Overheating
Firing incorrect 17
Fuel/air mixture incorrect 18
Lubrication inadequate 19
Miscellaneous causes 20

Clutch operating problems
Clutch slip 21
Clutch drag 22

Abnormal engine noise
Knocking or pinking 23

Piston slap or rattling from cylinder 24
Other noises 25

Abnormal transmission noise
Transmission noise 26

Exhaust smokes excessively
White/blue smoke (caused by oil burning) 27
Black smoke (caused by over-rich mixture) 28

Poor handling or roadholding
Directional instability 29
Steering bias to left or right 30
Handlebar vibrates or oscillates 31
Poor front fork performance 32
Front fork judder when braking 33
Poor rear suspension performance 34

Abnormal frame and suspension noise
Front end noise 35
Rear suspension noise 36

Brake problems
Brakes are spongy or ineffective 37
Brake drag 38
Brake lever pulsates in operation 39
Drum brake noise 40

Electrical problems
Battery dead or weak 41
Battery overcharged 42
Total electrical failure 43
Circuit failure 44
Bulbs blowing repeatedly 45

Starter motor problems
Starter motor not rotating 46
Starter motor rotates but engine does not turn over 47
Starter motor and clutch function but engine will not turn over ... 48

1 Introduction

This Section provides an easy reference-guide to the more common faults that are likely to afflict your machine. Obviously, the opportunities are almost limitless for faults to occur as a result of obscure failures, and to try and cover all eventualities would require a book. Indeed, a number have been written on the subject.

Successful fault diagnosis is not a mysterious 'black art' but the application of a bit of knowledge combined with a systematic and logical approach to the problem. Approach any fault diagnosis by first accurately identifying the symptom and then checking through the list of possibile causes, starting with the simplest or most obvious and progressing in stages to the most complex. Take nothing for granted, but above all apply liberal quantities of common sense.

The main symptom of a fault is given in the text as a major heading below which are listed, as Section headings, the various systems or areas which may contain the fault. Details of each possible cause for a fault and the remedial action to be taken are given, in brief, in the paragraphs below each Section heading. Further information should be sought in the relevant Chapter.

Engine does not start when turned over

2 No fuel flow to carburettor

● Fuel tank empty or level too low. If in doubt, prise off the fuel feed pipe at the carburettor end and check that fuel runs from pipe when the engine is cranked.
● Tank filler cap vent obstructed. This can prevent fuel from flowing into the carburettor float bowl because air cannot enter the fuel tank to replace it. The problem is more likely to appear when the machine is being ridden. Check by listening close to the filler cap and releasing it. A hissing noise indicates that a blockage is present. Remove the cap and clear the vent hole with wire or by using an air line from the inside of the cap.
● Fuel tap or filter blocked. Blockage may be due to accumulation of rust or paint flakes from the tank's inner surface or of foreign matter from contaminated fuel. Remove the tap and clean it and the filter. Look also for water droplets in the fuel.
● Fuel line blocked. Blockage of the fuel line is more likely to result from a kink in the line rather than the accumulation of debris.

3 Fuel not reaching cylinder

● Float chamber not filling. Caused by float needle or floats sticking in up position. This may occur after the machine has been left standing for an extended length of time allowing the fuel to evaporate. When this occurs a gummy residue is often left which hardens to a varnish-like substance. This condition may be worsened by corrosion and crystalline deposits produced prior to the total evaporation of contaminated fuel. Sticking of the float needle may also be caused by wear. In any case removal of the float chamber will be necessary for inspection and cleaning.
● Blockage in starting circuit, slow running circuit or jets. Blockage of these items may be attributable to debris from the fuel tank by-passing the filter system or to gumming up as described in paragraph 1. Water droplets in the fuel will also block jets and passages. The carburettor should be dismantled for cleaning.
● Fuel level too low. The fuel level in the float chamber is controlled by float height. The fuel level may increase with wear or damage but will never reduce, thus a low fuel level is an inherent rather than developing condition. Check the float height, renewing the float or needle if required.

4 Engine flooding

● Float valve needle worn or stuck open. A piece of rust or other debris can prevent correct seating of the needle against the valve seat thereby permitting an uncontrolled flow of fuel. Similarly, a worn needle or needle seat will prevent valve closure. Dismantle the carburettor float bowl for cleaning and, if necessary, renewal of the worn components.
● Fuel level too high. The fuel level is controlled by the float height which may increase due to wear of the float needle, pivot pin or operating tang. Check the float height, and make any necessary adjustments. A leaking float will cause an increase in fuel level, and thus should be renewed.
● Cold starting mechanism. Check the choke (starter mechanism) for correct operation. If the mechanism jams in the 'On' position subsequent starting of a hot engine will be difficult. Refer to Chapter 2 for details.
● Blocked air filter. A badly restricted air filter will cause flooding. Check the filter and clean or renew as required. A collapsed inlet hose will have a similar effect. Check that the air filter inlet has not become blocked by a rag or similar item.

5 No spark at plug

● Ignition switch not on.
● Engine stop switch off.
● Fuse blown. Check fuse for ignition circuit. See wiring diagram.
● Spark plug dirty, oiled or 'whiskered'. Because the induction mixture of a two-stroke engine is inclined to be of a rather oily nature it is comparatively easy to foul the plug electrodes, especially where there have been repeated attempts to start the engine. A machine used for short journeys will be more prone to fouling because the engine may never reach full operating temperature, and the deposits will not burn off. On rare occasions a change of plug grade may be required but the advice of a dealer should be sought before making such a change. 'Whiskering' is a comparatively rare occurrence on modern machines but may be encountered where pre-mixed petrol and oil (petroil) lubrication is employed. An electrode deposit in the form of a barely visible filament across the plug electrodes can short circuit the plug and prevent its sparking. On all two-stroke machines it is a sound precaution to carry a new spare spark plug for substitution in the event of fouling problems.
● Spark plug failure. Clean the spark plug thoroughly and reset the electrode gap. Refer to the spark plug section and the colour condition guide in Chapter 3. If the spark plug shorts internally or has sustained visible damage to the electrodes, core or ceramic insulator it should be renewed. On rare occasions a plug that appears to spark vigorously will fail to do so when refitted to the engine and subjected to the compression pressure in the cylinder.
● Spark plug cap or high tension (HT) lead faulty. Check condition and security. Replace if deterioration is evident. Most spark plug caps have an internal resistor designed to inhibit electrical interference with radio and television sets. On rare occasions the resistor may break down, thus preventing sparking. If this is suspected, fit a new cap as a precaution.
● Spark plug cap loose. Check that the spark plug cap fits securely over the plug and, where fitted, the screwed terminal on the plug end is secure.
● Shorting due to moisture. Certain parts of the ignition system are susceptible to shorting when the machine is ridden or parked in wet weather. Check particularly the area from the spark plug cap back to the ignition coil. A water dispersant spray may be used to dry out waterlogged components. Recurrence of the problem can be prevented by using an ignition sealant spray after drying out and cleaning.
● Ignition or stop switch shorted. May be caused by water corrosion or wear. Water dispersant and contact cleaning sprays may be used. If this fails to overcome the problem dismantling and visual inspection of the switches will be required.
● Shorting or open circuit in wiring. Failure in any wire connecting any of the ignition components will cause ignition malfunction. Check also that all connections are clean, dry and tight.
● Ignition coil failure. Check the coil, referring to Chapter 3.
● Fault in CDI system. Refer to Chapter 3 for detailed fault diagnosis procedure. Test and renew system components as required.

6 Weak spark at plug

● Feeble sparking at the plug may be caused by any of the faults mentioned in the preceding Section other than those items in the first three paragraphs. Check first the spark plug, this being the most likely culprit.

7 Compression low

● Spark plug loose. This will be self-evident on inspection, and may be accompanied by a hissing noise when the engine is turned over. Remove the plug and check that the threads in the cylinder head are not damaged. Check also that the plug sealing washer is in good condition.
● Cylinder head gasket leaking. This condition is often accompanied by a high pitched squeak from around the cylinder head and oil loss, and may be caused by insufficiently tightened cylinder head fasteners, a warped cylinder head or mechanical failure of the gasket material. Re-torqueing the fasteners to the correct specification may seal the leak in some instances but if damage has occurred this course of action will provide, at best, only a temporary cure.
● Low crankcase compression. This can be caused by worn main bearings and seals and will upset the incoming fuel/air mixture. A good seal in these areas is essential on any two-stroke engine.
● Piston rings sticking or broken. Sticking of the piston rings may be caused by seizure due to lack of lubrication or overheating as a result of poor carburation or incorrect fuel type. Gumming of the rings may result from lack of use, or carbon deposits in the ring grooves. Broken rings result from over-revving, over-heating or general wear. In either case a top-end overhaul will be required.

Engine stalls after starting

8 General causes

● Improper cold start mechanism operation. Refer to Chapter 2 for details of test procedures. A cold engine may not require application of an enriched mixture to start initially but may baulk without choke once firing. Likewise a hot engine may start with an enriched mixture but will stop almost immediately if the choke is inadvertently in operation.
● Ignition malfunction. See Section 9. Weak spark at plug.
● Carburettor incorrectly adjusted. Maladjustment of the mixture strength or idle speed may cause the engine to stop immediately after starting. See Chapter 2.
● Fuel contamination. Check for filter blockage by debris or water which reduces, but does not completely stop, fuel flow, or blockage of the slow speed circuit in the carburettor by the same agents. If water is present it can often be seen as droplets in the bottom of the float bowl. Clean the filter and, where water is in evidence, drain and flush the fuel tank and float bowl.
● Intake air leak. Check for security of the carburettor mounting and hose connections, and for cracks or splits in the hoses. Check also that the carburettor top is secure and that the vacuum gauge adaptor plug (where fitted) is tight.
● Air filter blocked or omitted. A blocked filter will cause an over-rich mixture; the omission of a filter will cause an excessively weak mixture. Both conditions will have a detrimental effect on carburation. Clean or renew the filter as necessary.
● Fuel filler cap air vent blocked. Usually caused by dirt or water. Clean the vent orifice.
● Choked exhaust system. Caused by excessive carbon build-up in the system, particularly around the silencer baffles. In many cases these can be detached for cleaning, though mopeds have one-piece systems which require a rather different approach. Refer to Chapter 2 for further information.
● Excessive carbon build-up in the engine. This can result from failure to decarbonise the engine at the specified interval or through excessive oil consumption. Check pump adjustment

Poor running at idle and low speed

9 Weak spark at plug or erratic firing

● Battery voltage low. In certain conditions low battery charge, especially when coupled with a badly sulphated battery, may result in misfiring. If the battery is in good general condition it should be recharged; an old battery suffering from sulphated plates should be renewed.
● Spark plug fouled, faulty or incorrectly adjusted. See Section 4 or refer to Chapter 3.
● Spark plug cap or high tension lead shorting. Check the condition of both these items ensuring that they are in good condition and dry and that the cap is fitted correctly.
● Spark plug type incorrect. Fit plug of correct type and heat range as given in Specifications. In certain conditions a plug of hotter or colder type may be required for normal running.
● Ignition timing incorrect. Check the ignition timing dynamically, ensuring that the advance is functioning correctly. If incorrect, check the CDI unit.
● Faulty ignition coil. Partial failure of the coil internal insulation will diminish the performance of the coil. No repair is possible, a new component must be fitted.
● Defective flywheel generator ignition source. Refer to Chapter 3 for further details on test procedures.

10 Fuel/air mixture incorrect

● Intake air leak. Check carburettor mountings and air cleaner hoses for security and signs of splitting. Ensure that carburettor top is tight.
● Mixture strength incorrect. Adjust slow running mixture strength using pilot adjustment screw.
● Pilot jet or slow running circuit blocked. The carburettor should be removed and dismantled for thorough cleaning. Blow through all jets and air passages with compressed air to clear obstructions.
● Air cleaner clogged or omitted. Clean or fit air cleaner element as necessary. Check also that the element and air filter cover are correctly seated.
● Cold start mechanism in operation.
● Fuel level too high or too low. Check the float height, renewing float or needle if required. See Section 3 or 4.
● Fuel tank air vent obstructed. Obstructions usually caused by dirt or water. Clean vent orifice.

11 Compression low

● See Section 7.

Acceleration poor

12 General causes

● All items as for previous Section.
● Choked air filter. Failure to keep the air filter element clean will allow the build-up of dirt with proportional loss of performance. In extreme cases of neglect acceleration will suffer.
● Choked exhaust system. This can result from failure to remove accumulations of carbon from the silencer baffles at the prescribed intervals. The increased back pressure will make the machine noticeably sluggish. Refer to Chapter 2 for further information on decarbonisation.
● Excessive carbon build-up in the engine. This can result from failure to decarbonise the engine at the specified interval or through excessive oil consumption. Check pump adjustment.
● Ignition timing incorrect. Check the ignition timing as described in Chapter 3. Where no provision for adjustment exists, test the electronic ignition components and renew as required.
● Carburation fault. See Section 10.
● Mechanical resistance. Check that the brakes are not binding. On

small machines in particular note that the increased rolling resistance caused by under-inflated tyres may impede acceleration.

● Transmission defect. Check belt, clutch and pulleys for wear or damage, referring to Routine Maintenance for details.

Poor running or lack of power at high speeds

13 Weak spark at plug or erratic firing

● All items as for Section 9.
● HT lead insulation failure. Insulation failure of the HT lead and spark plug cap due to old age or damage can cause shorting when the engine is driven hard. This condition may be less noticeable, or not noticeable at all at lower engine speeds.

14 Fuel/air mixture incorrect

● All items as for Section 10, with the exception of items relative exclusively to low speed running.
● Main jet blocked. Debris from contaminated fuel, or from the fuel tank, and water in the fuel can block the main jet. Clean the fuel filter, the float bowl area, and if water is present, flush and refill the fuel tank.
● Jet needle and needle jet worn. These can be renewed individually but should be renewed as a pair. Renewal of both items requires partial dismantling of the carburettor.
● Air bleed holes blocked. Dismantle carburettor and use compressed air to blow out all air passages.
● Reduced fuel flow. A reduction in the maximum fuel flow from the fuel tank to the carburettor will cause fuel starvation, proportionate to the engine speed. Check for blockages through debris or a kinked fuel line.

15 Compression low

● See Section 7.

Knocking or pinking

16 General causes

● Carbon build-up in combustion chamber. After high mileages have been covered large accumulations of carbon may occur. These may glow red hot and cause premature ignition of the fuel/air mixture, in advance of normal firing by the spark plug. Cylinder head removal will be required to allow inspection and cleaning.
● Fuel incorrect. A low grade fuel, or one of poor quality may result in compression induced detonation of the fuel resulting in knocking and pinking noises. Old fuel can cause similar problems. A too highly leaded fuel will reduce detonation but will accelerate deposit formation in the combustion chamber and may lead to early pre-ignition as described in item 1.
● Spark plug heat range incorrect. Uncontrolled pre-ignition can result from the use of a spark plug the heat range of which is too hot.
● Weak mixture. Overheating of the engine due to a weak mixture can result in pre-ignition occurring where it would not occur when engine temperature was within normal limits. Maladjustment, blocked jets or passages and air leaks can cause this condition.

Overheating

17 Firing incorrect

● Spark plug fouled, defective or maladjusted. See Section 5.
● Spark plug type incorrect. Refer to the Specifications and ensure that the correct plug type is fitted.

18 Fuel/air mixture incorrect

● Slow speed mixture strength incorrect. Adjust pilot air screw.
● Main jet wrong size. The carburettor is jetted for sea level atmospheric conditions. For high altitudes, usually above 5000 ft, a smaller main jet will be required.
● Air filter badly fitted or omitted. Check that the filter element is in place and that it and the air filter box cover are sealing correctly. Any leaks will cause a weak mixture.
● Induction air leaks. Check the security of the carburettor mountings and hose connections, and for cracks and splits in the hoses. Check also that the carburettor top is secure.
● Fuel level too low. See Section 3.
● Fuel tank filler cap air vent obstructed. Clear blockage.

19 Lubrication inadequate

● Oil tank empty or low. This will have disastrous consequences if left unnoticed. Check and replenish the tank regularly.

20 Miscellaneous causes

● Engine fins clogged. A build-up of mud in the cylinder head and cylinder barrel cooling fins will decrease the cooling capabilities of the fins. Clean the fins as required.

Clutch operating problems

21 Clutch slip

● Clutch shoes contaminated with oil. Rectify oil leak and fit new shoes.
● Clutch shoes glazed. Remove glazing with abrasive paper, or renew shoes.
● Clutch shoes worn or friction material separated from shoe(s).

22 Clutch drag

● Worn clutch shoe return springs. Dismantle and renew springs.

Abnormal engine noise

23 Knocking or pinking

● See Section 16.

24 Piston slap or rattling from cylinder

● Cylinder bore/piston clearance excessive. Resulting from wear, or partial seizure. This condition can often be heard as a high, rapid tapping noise when the engine is under little or no load, particularly when power is just beginning to be applied. Reboring to the next correct oversize should be carried out and a new oversize piston fitted.
● Connecting rod bent. This can be caused by over-revving, trying to start a very badly flooded engine (resulting in a hydraulic lock in the cylinder) or by earlier mechanical failure. Attempts at straightening a bent connecting rod from a high performance engine are not recommended. Careful inspection of the crankshaft should be made before renewing the damaged connecting rod.
● Gudgeon pin, piston boss bore or small-end bearing wear or seizure. Excess clearance or partial seizure between normal moving parts of these items can cause continuous or intermittent tapping noises. Rapid wear or seizure is caused by lubrication starvation.
● Piston rings worn, broken or sticking. Renew the rings after careful inspection of the piston and bore.

25 Other noises

● Big-end bearing wear. A pronounced knock from within the crankcase which worsens rapidly is indicative of big-end bearing failure as a result of extreme normal wear or lubrication failure. Remedial action in the form of a bottom end overhaul should be taken; continuing to run the engine will lead to further damage including the possibility of connecting rod breakage.

● Main bearing failure. Extreme normal wear or failure of the main bearings is characteristically accompanied by a rumble from the crankcase and vibration felt through the frame and footrests. Renew the worn bearings and carry out a very careful examination of the crankshaft.

● Crankshaft excessively out of true. A bent crank may result from over-revving or damage from an upper cylinder component or gearbox failure. Damage can also result from dropping the machine on either crankshaft end. Straightening of the crankshaft is not possible in normal circumstances; a replacement item should be fitted.

● Engine mounting loose. Tighten all the engine mounting nuts and bolts.

● Cylinder head gasket leaking. The noise most often associated with a leaking head gasket is a high pitched squeaking, although any other noise consistent with gas being forced out under pressure from a small orifice can also be emitted. Gasket leakage is often accompanied by oil seepage from around the mating joint or from the cylinder head holding down bolts and nuts. Leakage results from insufficient or uneven tightening of the cylinder head fasteners, or from random mechanical failure. Retightening to the correct torque figure will, at best, only provide a temporary cure. The gasket should be renewed at the earliest opportunity.

● Exhaust system leakage. Popping or crackling in the exhaust system, particularly when it occurs with the engine on the overrun, indicates a poor joint at the cylinder port. Failure of the gasket or looseness of the clamp should be looked for.

Abnormal transmission noise

26 Transmission noise

● Bearing or bushes worn or damaged. Renew the affected components.

● Gear pinions worn or chipped. Renew the gear pinions.

● Metal chips jammed in gear teeth. This can occur when pieces of metal from any failed component are picked up by a meshing pinion. The condition will lead to rapid bearing wear or early gear failure.

● Transmission oil level too low. Top up immediately to prevent damage to gearbox.

● Damaged or slipping drive belt.

● Worn or damaged driving pulley mechanism.

● Worn or damaged clutch/driven pulley.

Exhaust smokes excessively

27 White/blue smoke (caused by oil burning)

● Piston rings worn or broken. Breakage or wear of any ring, but particularly the oil control ring, will allow engine oil past the piston into the combustion chamber. Examine and renew, where necessary, the cylinder barrel and piston.

● Cylinder cracked, worn or scored. These conditions may be caused by overheating, lack of lubrication, component failure or advanced normal wear. The cylinder barrel should be renewed and, if necessary, a new piston fitted.

● Oil pump operation incorrect. Check the oil pump as described in Chapter 2.

● Accumulated oil deposits in exhaust system. If the machine is used for short journeys only it is possible for the oil residue in the exhaust gases to condense in the relatively cool silencer. If the machine is then taken for a longer run in hot weather, the accumulated oil will burn off producing ominous smoke from the exhaust.

28 Black smoke (caused by over-rich mixture)

● Air filter element clogged. Clean or renew the element.

● Main jet loose. Remove the float chamber to check for tightness of the jet.

● Cold start mechanism jammed on. Check that the mechanism works smoothly and correctly.

● Fuel level too high. The fuel level is controlled by the float height which can increase as a result of wear or damage. Remove the float bowl and check the float height. Check also that floats have not punctured; a punctured float will lose buoyancy and allow an increased fuel level.

● Float valve needle stuck open. Caused by dirt or a worn valve. Clean the float chamber or renew the needle and, if necessary, the valve seat.

Poor handling or roadholding

29 Directional instability

● Steering head bearing adjustment too tight. This will cause rolling or weaving at low speeds. Re-adjust the bearings.

● Steering head bearing worn or damaged. Correct adjustment of the bearing will prove impossible to achieve if wear or damage has occurred. Inconsistent handling will occur including rolling or weaving at low speed and poor directional control at indeterminate higher speeds. The steering head bearing should be dismantled for inspection and renewed if required. Lubrication should also be carried out.

● Bearing races pitted or dented. Impact damage caused, perhaps, by an accident or riding over a pot-hole can cause indentation of the bearing, usually in one position. This should be noted as notchiness when the handlebars are turned. Renew and lubricate the bearings.

● Steering stem bent. This will occur only if the machine is subjected to a high impact such as hitting a curb or a pot-hole. The stem should be renewed; do not attempt to straighten it.

● Front or rear tyre pressures too low.

● Front or rear tyre worn. General instability, high speed wobbles and skipping over white lines indicates that tyre renewal may be required. Tyre induced problems, in some machine/tyre combinations, can occur even when the tyre in question is by no means fully worn.

● Rear suspension pivot bushes worn. Difficulties in holding line, particularly when cornering or when changing power settings indicates wear in the pivot bushes. The engine/transmission unit should be removed from the machine and the bushes renewed.

● Wheel bearings worn. Renew the worn bearings.

● Tyres unsuitable for machine. Not all available tyres will suit the characteristics of the frame and suspension, indeed, some tyres or tyre combinations may cause a transformation in the handling characteristics. If handling problems occur immediately after changing to a new tyre type or make, revert to the original tyres to see whether an improvement can be noted. In some instances a change to what are, in fact, suitable tyres may give rise to handling deficiences. In this case a thorough check should be made of all frame and suspension items which affect stability.

30 Steering bias to left or right

● Wheels out of alignment. This can be caused by impact damage to the frame, swinging arm, wheel spindles or front forks. Although occasionally a result of material failure or corrosion it is usually as a result of a crash.

31 Handlebar vibrates or oscillates

● Tyres worn or out of balance. Either condition, particularly in the front tyre, will promote shaking of the fork assembly and thus the handlebars. A sudden onset of shaking can result if a balance weight is displaced during use.

● Tyres badly positioned on the wheel rims. A moulded line on each wall of a tyre is provided to allow visual verification that the tyre is correctly positioned on the rim. A check can be made by rotating the tyre; any misalignment will be immediately obvious.

● Wheel rims warped or damaged. Inspect the wheels for runout as described in Chapter 5.

● Wheel bearings worn. Renew the bearings.

● Steering head bearings incorrectly adjusted. Vibration is more likely to result from bearings which are too loose rather than too tight. Re-adjust the bearings.

● Loosen fork component fasteners. Loose nuts and bolts holding the fork legs, wheel spindle, mudguards or steering stem can promote shaking at the handlebars. Fasteners on running gear such as the forks and suspension should be check tightened occasionally to prevent dangerous looseness of components occurring.

● Engine mounting bolts loose. Tighten all fasteners.

32 Poor front fork performance

● Weak fork springs. Progressive fatigue of the fork springs, resulting in a reduced spring free length, will occur after extensive use. This condition will promote excessive fork dive under braking, and in its advanced form will reduce the at-rest extended length of the forks and thus the fork geometry. Renewal of the springs as a pair is the only satisfactory course of action.

● Suspension units leaking or damaged. Renew as a pair.

33 Front fork judder when braking

● Slack steering head bearings. Re-adjust the bearings.

● Warped brake drum. If irregular braking action occurs fork judder can be induced in what are normally serviceable forks. Renew the damaged brake components.

34 Poor rear suspension performance

● Rear suspension unit damper worn out or leaking. The damping performance of most rear suspension units falls off with age. This is a gradual process, and thus may not be immediately obvious. Indications of poor damping include hopping of the rear end when cornering or braking, and a general loss of positive stability.

● Weak rear spring. If the suspension unit spring fatigues it will promote excessive pitching of the machine and reduce the ground clearance when cornering. Although replacement springs are available separately from the rear suspension damper unit it is probable that if spring fatigue has occurred the damper unit will also require renewal.

● Bent suspension unit damper rod. This is likely to occur only if the machine is dropped or if seizure of the piston occurs. If either happens the suspension unit should be renewed.

Abnormal frame and suspension noise

35 Front end noise

● Spring weak or broken. Makes a clicking or scraping sound.

● Steering head bearings loose or damaged. Clicks when braking. Check, adjust or replace. Make sure all fork clamp pinch bolts are tight.

● Fork clamps loose. Make sure all fork clamp pinch bolts are tight.

36 Rear suspension noise

● Leakage of a suspension unit, usually evident by oil on the outer surfaces, can cause a spurting noise. The suspension unit should be renewed.

● Defective rear suspension unit with internal damage. Renew the suspension unit.

Brake problems

37 Brakes are spongy or ineffective

● Brake cable deterioration. Damage to the outer cable by stretching or being trapped will give a spongy feel to the brake lever. The cable should be renewed. A cable which has become corroded due to old age or neglect of lubrication will partially seize making operation very heavy. Lubrication at this stage may overcome the problem but the fitting of a new cable is recommended.

● Worn brake linings. Determine lining wear using the external brake wear indicator on the brake backplate, or by removing the wheel and withdrawing the brake backplate. Renew the shoe/lining units as a pair if the linings are worn below the recommended limit.

● Worn brake camshaft. Wear between the camshaft and the bearing surface will reduce brake feel and reduce operating efficiency. Renewal of one or both items will be required to rectify the fault.

● Worn brake cam and shoe ends. Renew the worn components.

● Linings contaminated with dust or grease. Any accumulations of dust should be cleaned from the brake assembly and drum using a petrol dampened cloth. Do not blow or brush off the dust because it is asbestos based and thus harmful if inhaled. Light contamination from grease can be removed from the surface of the brake linings using a solvent; attempts at removing heavier contamination are less likely to be successful because some of the lubricant will have been absorbed by the lining material which will severely reduce the braking performance.

38 Brake drag

● Incorrect adjustment. Re-adjust the brake operating mechanism.

● Drum warped or oval. This can result from overheating or impact or uneven tension of the wheel spokes. The condition is difficult to correct, although if slight ovality only occurs, skimming the surface of the brake drum can provide a cure. This is work for a specialist engineer. Renewal of the complete wheel hub is normally the only satisfactory solution.

● Weak brake shoe return springs. This will prevent the brake lining/shoe units from pulling away from the drum surface once the brake is released. The springs should be renewed.

● Brake camshaft, lever pivot or cable poorly lubricated. Failure to attend to regular lubrication of these areas will increase operating resistance which, when compounded, may cause tardy operation and poor release movement.

39 Brake lever pulsates in operation

● Drums warped or oval. This can result from overheating or impact. This condition is difficult to correct, although if slight ovality only occurs skimming the surface of the drum can provide a cure. This is work for a specialist engineer. Renewal of the hub is normally the only satisfactory solution.

40 Drum brake noise

● Drum warped or oval. This can cause intermittent rubbing of the brake linings against the drum. See the preceding Section.

● Brake linings glazed. This condition, usually accompanied by heavy lining dust contamination, often induces brake squeal. The surface of the linings may be roughened using glass-paper or a fine file.

Electrical problems

41 Battery dead or weak

● Battery faulty. Battery life should not be expected to exceed 3 to 4 years, particularly where a starter motor is used regularly. Gradual sulphation of the plates and sediment deposits will reduce the battery performance. Plate and insulator damage can often occur as a result of vibration. Complete power failure, or intermittent failure, may be due to a broken battery terminal.

● Battery leads making poor contact. Remove the battery leads and clean them and the terminals, removing all traces of corrosion and tarnish. Reconnect the leads and apply a coating of petroleum jelly to the terminals.

● Load excessive. If additional items such as spot lamps, are fitted, which increase the total electrical load above the maximum alternator output, the battery will fail to maintain full charge. Reduce the electrical load to suit the electrical capacity.

● Rectifier failure.

● Alternator generating coils open-circuit or shorted.

● Charging circuit shorting or open circuit. This may be caused by frayed or broken wiring, dirty connectors or a faulty ignition switch. The system should be tested in a logical manner. See Section 54.

42 Battery overcharged

● Regulator or ballast resistor faulty. Overcharging is indicated if the battery becomes hot or it is noticed that the electrolyte level falls repeatedly between checks.

● Battery wrongly matched to the electrical circuit. Ensure that the specified battery is fitted to the machine.

43 Total electrical failure

● Fuse blown. Check the main fuse. If a fault has occurred, it must be rectified before a new fuse is fitted.

● Battery faulty. See Section 51.

● Earth failure. Check that the frame main earth strap from the battery is securely affixed to the frame and is making a good contact.

● Ignition switch or power circuit failure. Check for current flow through the battery positive lead (red) to the ignition switch. Check the ignition switch for continuity.

44 Circuit failure

● Cable failure. Refer to the machine's wiring diagram and check the circuit for continuity. Open circuits are a result of loose or corroded connections, either at terminals or in-line connectors, or because of broken wires. Occasionally, the core of a wire will break without there being any apparent damage to the outer plastic cover.

● Switch failure. All switches may be checked for continuity in each switch position, after referring to the switch position boxes incorporated in the wiring diagram for the machine. Switch failure may be a result of mechanical breakage, corrosion or water.

● Fuse blown. Refer to the wiring diagram to check whether or not a circuit fuse is fitted. Replace the fuse, if blown, only after the fault has been identified and rectified.

45 Bulbs blowing repeatedly

● Vibration failure. This is often an inherent fault related to the natural vibration characteristics of the engine and frame and is, thus, difficult to resolve. Modifications of the lamp mounting, to change the damping characteristics, may help.

● Intermittent earth. Repeated failure of one bulb, particularly where the bulb is fed directly from the generator, indicates that a poor earth exists somewhere in the circuit. Check that a good contact is available at each earthing point in the circuit.

Starter motor problems

46 Starter motor not rotating

● Engine stop switch off.

● Fuse blown. Check the main fuse located behind the battery side cover.

● Battery voltage low. Switching on the turn signals and operating the horn will give a good indication of the charge level. If necessary recharge the battery from an external source.

● Rear brake not locked on, interlock switch off.

● Ignition switch defective. Check switch for continuity and connections for security.

● Starter button switch faulty. Check continuity of switch. Faults as for engine stop switch.

● Starter relay (solenoid) faulty. If the switch is functioning correctly a pronounced click should be heard when the starter button is depressed. This presupposes that current is flowing to the solenoid when the button is depressed.

● Wiring open or shorted. Check first that the battery terminal connections are tight and corrosion free. Follow this by checking that all wiring connections are dry, tight and corrosion free. Check also for frayed or broken wiring. Occasionally a wire may become trapped between two moving components, particularly in the vicinity of the steering head, leading to breakage of the internal core but leaving the softer but more resilient outer cover intact. This can cause mysterious intermittent or total power loss.

● Starter motor defective. A badly worn starter motor may cause high current drain from a battery without the motor rotating. If current is found to be reaching the motor, after checking the starter button and starter relay, suspect a damaged motor. The motor should be removed for inspection.

47 Starter motor rotates but engine does not turn over

● Damaged starter motor drive train. Inspect and renew component where necessary. Failure in this area is unlikely.

48 Starter motor and clutch function but engine will not turn over

● Engine seized. Seizure of the engine is always a result of damage to internal components due to lubrication failure, or component breakage resulting from abuse, neglect or old age. A seizing or partially seized component may go un-noticed until the engine has cooled down and an attempt is made to restart the engine. Instantaneous seizure whilst the engine is running indicates component breakage. In either case major dismantling and inspection will be required.

Routine maintenance

Refer to Chapter 7 for information relating to the SA50 Met-in

Specifications

Engine/transmission

Spark plug gap	0.6 – 0.7 mm (0.024 – 0.028 in)
Engine idle speed	1800 ± 100 rpm
Throttle free play	2 – 6 mm ($1/8$ – $1/4$ in) at edge of throttle twistgrip flange

Frame/suspension

Brake lever free play (front and rear)	10 – 20 mm ($3/8$ – $3/4$ in)
Tyre pressures (cold):	
Front	21 psi (1.5 kg/cm²)
Rear	28 psi (2.0 kg/cm²)

Recommended lubricants

Engine:	
Oil tank capacity	0.8 litre (1.4 Imp pint)
Oil grade	Honda 2-stroke injector oil, or equivalent
Transmission final reduction gearbox:	
Oil grade	SAE 10W/40 motor oil
Oil capacity...	90 cc (3.2 Imp fl oz)
Pivots and bearings	General purpose grease
Control cables	Oil or cable lubricant
Speedometer cable	Grease or cable lubricant
Speedometer drive	General purpose grease

Introduction

Periodic routine maintenance is a continuous process which should commence immediately the machine is used. The object is to maintain all adjustments and to diagnose and rectify minor defects before they develop into more extensive, and often more expensive, problems.

It follows that if the machine is maintained properly, it will both run and perform with optimum efficiency and be less prone to unexpected breakdowns. Regular inspection of the machine will show up any parts which are wearing, and with a little experience, it is possible to obtain the maximum life from any one component, renewing it when it becomes so worn that it is liable to fail.

Regular cleaning can be considered as important as mechanical maintenance. This will ensure that all the cycle parts are inspected regularly and are kept free from accumulations of road dirt and grime.

Cleaning is especially important during the winter months, despite its appearance of being a thankless task which very soon seems pointless. On the contrary, it is during these months that the paintwork, chromium plating, and the alloy casings suffer the ravages of abrasive grit, rain and road salt. A couple of hours spent weekly on cleaning the machine will maintain its appearance and value, and highlight small points, like chipped paint, before they become a serious problem.

The various maintenance tasks are described under their respective mileage and calendar headings, and are accompanied by diagrams and photographs where pertinent.

It should be noted that the intervals between each maintenance task serve only as a guide. As the machine gets older, or if it is used under particularly arduous conditions, it is advisable to reduce the period between each check.

For ease of reference, most service operations are described in detail under the relevant heading. However, if further general information is required, this can be found under the pertinent Section heading and Chapter in the main text.

Although no special tools are required for routine maintenance, a good selection of general workshop tools is essential. Included in the tools must be a range of metric ring or combination spanners, a selection of crosshead screwdrivers, and two pairs of circlip pliers, one external opening and the other internal opening. Additionally, owing to the extreme tightness of most casing screws, an impact screwdriver, together with a choice of large or small cross-head screw bits, is absolutely indispensable. This is particularly so if the engine has not been dismantled since leaving the factory.

Pre-ride checks

The following check list should be followed prior to riding the machine. With a little practice they should become second nature and will take only a few minutes to carry out. Although simple, the checks serve to alert the rider of any potentially dangerous or inconvenient faults before setting off.

1 Check the engine oil level

An LED warning indicator in the instrument panel serves to warn the rider of a low oil level in the tank. At this point the tank must be replenished with the recommended grade of oil by lifting the seat, removing the filler cap from the oil tank and topping up to the upper level mark on the tank.

It is good practice, however, to regularly check the oil level visually, through the translucent oil tank. This will serve as a double-check, should the level checking circuit fail, and will also ensure that the oil does not reach an unsafe level during a journey.

Note: If the low oil level warning lamp comes on when riding the machine ensure that the oil tank is replenished without delay. In the event of the oil level falling low enough for air to enter the pipe between the oil tank and engine, the system must be bled before the machine is ridden. The bleeding operation is described in Chapter 2.

2 Check the fuel level

With the ignition switch on, note the position of the fuel gauge needle. If the needle is on or near the red E sector, the tank should be filled at the earliest opportunity. A full tank contains about 3.7 litre (0.81 Imp gallon). When reading empty (E) about 1.0 litre (1.6 Imp pint) remains. Refill the tank with regular or low-lead petrol with a Research Octane number of 91 or higher (2-Star).

3 Check the lights and turn signals

Check the operation of the turn signals, the turn signal warning lamp, the brake lamp (front and rear brakes) and the horn (ignition switch on).

4 Check the tyres

Examine both tyres for signs of damage, including splits or bulges in the sidewalls, and low pressure.

5 Check the operation of the brakes

With the machine off its centre stand roll it forward and check that the front and rear brakes operate correctly.

6 Check the steering

Check that the steering moves smoothly and easily from lock to lock.

7 Check the throttle operation

Check that the throttle twistgrip operates normally and that it returns to the closed position when released. **DO NOT** carry out this check with the engine running; the automatic transmission may engage, causing the machine to move off without you!

8 Check the lights and controls

Operate the rear brake lever and engage the lock to hold it on. Switch on the ignition and press the starter. Leave the engine to warm up while checking that the headlamp and tail lamp are working and that the instrument panel illumination lamps come on. Check that the rear view mirrors are adjusted correctly and that any luggage is properly secured on the carrier.

Weekly, or every 50 miles (75 km)

1 Legal check

Check the operation of the electrical system, ensuring that the lights and horn are working properly and that the lenses are clean. Note that in the UK it is an offence to use a vehicle on which the lights are defective. This applies even when the machine is used during daylight hours. The horn is also a statutory requirement.

Inspect each tyre to ensure that its tread is visible over the entire circumference with no bald patches.

2 Safety check

Give the machine a close visual inspection, checking for loose nuts and fittings, frayed control cables etc. Check the tyres for damage, especially splitting on the sidewalls. Remove any stones or other objects caught between the treads. This is particularly important on the front tyre, where rapid deflation due to penetration of the inner tube will almost certainly cause total loss of control.

3 Tyre pressures

Check the tyre pressures. Always check with the tyres cold, using a pressure gauge known to be accurate. It is recommended that a pocket pressure gauge is purchased to offset any fluctuation between garage forecourt instruments. Refer to the recommended pressures given below and the tyre information label on the bodywork centre panel.

Front .... 21 psi (1.50 kg/cm²)
Rear .... 28 psi (2.00 kg/cm²)

4 Control cable lubrication

Apply a few drops of oil to the exposed lengths of inner cable at the tops of the various control cables. This will prevent the cables drying out between the more thorough lubrication given during the yearly/2000 mile service.

6 monthly, or every 1000 miles (1500 km)

1 Removing and refitting body panels
General

To gain access to most of the service items on the machine, it is first necessary to remove one or more of the body panels. To avoid repetition, the procedure for removing each panel is described below. Where a particular operation requires that a panel be removed before work commences, this is mentioned in the text, and this section should be consulted for details.

The various panels are of plastic construction and should be handled carefully. Whilst fairly strong, the surface finish is easily damaged if handled carelessly, and the locating tabs in particular are easily broken. Never force any panel; check that all of the fasteners have been removed and that removal is being tackled in the right sequence. Once detached, place the panel on some soft cloth well away from the working area. Keep all panels away from brake fluid, solvents and thinners, any of which will damage the surface.

Left-hand side panel

Unlock and open the seat. Remove the two bolts and two domed nuts which secure the rear carrier and lift it away, taking care not to scratch the bodywork. On the NB50 Vision-X model, take care not to lose the two spacers fitted to the carrier mounting studs. Release the single domed nut below the front of the seat and the two screws at the lower edge of the left-hand side panel. Carefully remove the side panel and place it to one side.

In the case of the NE50 Vision model, remove the carrier as detailed above, then remove the single domed nut below the front of the seat, the screw immediately below it, and the single screw at the front lower edge of the side panel. Disengage the panel at the rear, taking care to avoid damaging the locating tabs, then swing it upwards and outwards until it can be freed at the front lower corner.

Right-hand side panel

Remove the left-hand side panel as detailed above. On the NB50 Vision-X model, open the luggage locker in the right-hand side panel and remove the single recessed bolt just below the latch plate. On both models, remove the screw(s) at the lower edge of the panel and lift it clear of the frame.

Front cover

The legshield assembly is a double-skinned structure comprising the front cover and the legshield mouldings. These are held together by ten screws fitted from the inner edge of the legshield. To release the front cover, remove the ten screws securing it to the legshield and then release the three domed nuts which secure the cover to the frame. The cover can then be lifted away exposing the legshield and much of the wiring.

Floorboard

Remove the left-hand and right-hand side panels as described above. Unscrew the four bolts which retain the floorboard to the outrigger brackets beneath it and manoeuvre it clear of the frame.

Legshield

Remove the side panels, floorboard and front cover as detailed above. Release the turn signal relay from its bracket on the legshield and remove the single screw to free the multi-pin connector. On later models disconnect the front turn signal wiring.

Moving to the inside of the legshield, release the two screws which secure its lower edge to the frame outriggers. Unlock the lid of the front luggage compartment and remove the single retaining bolt. The legshield can now be lifted away and placed to one side.

Reassembly

In most respects, the various body panels are refitted by reversing the removal sequence. Note that this extends to the order of removal, ie the side panels and floorboards should not be fitted until the legshield is in place. Make sure that no undue force is used during installation and that any spacers and washers are refitted correctly to avoid straining the panels. Be particularly careful to ensure that all locating tabs are positioned correctly so that they hold the panel edges in alignment. It will be obvious from examination if they are misaligned.

2 Cleaning the air filter element

This operation should be carried out without fail at this interval, and more frequently if the machine is used in particularly dusty conditions. If the element is allowed to become clogged with dust, engine performance and fuel consumption will suffer, and if the element is damaged or missing, rapid engine wear can be expected. Start by removing the left-hand side panel as described under the previous heading.

Release the four screws which retain the air filter cover and lift it away. The flat foam element can now be lifted out of the casing for cleaning. Honda caution against the use of petrol for cleaning the element because of the high fire risk, recommending instead a high flash-point or non-flammable solvent. The most easily obtained solvent of this type is methylated spirit, available from most hardware and chemist shops. Wash the element thoroughly, then squeeze out any excess solvent and allow it to dry. Do not wring the element out because this will damage the foam. Check the element carefully for holes or splits and renew it if damaged in any way. Soak the cleaned or new element in SAE 10W/40 motor oil, then squeeze it to leave the surface moist, but not dripping. Refit the element and install the cover.

3 Idle speed adjustment

The engine idle speed must be set correctly or problems will arise; if too high the machine will tend to 'creep' when at a standstill, whilst if it is too low the engine may stall repeatedly. If neither problem is evident, the idle speed check can be ignored for most practical purposes.

To set the idle speed accurately, a tachometer will be required; a piece of equipment not available to most owners. In practice it is possible to perform this task with reasonable accuracy without the use of a tachometer, provided that care is taken, but if in doubt, leave this job to a Honda dealer.

Some precautions must be taken when carrying out the adjustment. Start by warming up the engine to its normal operating temperature by riding it for a few miles, then place it on its centre stand on level ground. Make sure that the machine is secure and that the rear wheel is well clear of any tools or other obstructions. It will spin during the idle speed check and thus presents a hazard to the owner if care is not taken.

Start the engine and release the rear brake lock. The rear wheel will tend to turn while the engine is running due to the automatic transmission design, but if when the rear brake is applied the engine slows noticeably, this indicates that the idle speed is too high and that the transmission is partially engaged and thus 'dragging' a little.

The idle speed screw is located on the left-hand side of the carburettor and can be reached with the side panel in place if required, although it is easier to find if the panel is removed. On the NB50 Vision-X model there are two small holes in the side panel, and the idle speed screw can be turned by passing a screwdriver blade through the upper of the two holes. In the case of the NE50 Vision model, the panel is cut away and does not cover the screws. On both models, do not disturb the pilot screw which is situated below and forward of the idle speed screw.

If a tachometer is available, set the idle speed to 1800 ± 100 rpm. In the absence of a tachometer, set the idle speed screw to obtain the lowest possible reliable idle speed. Once set, the rear wheel should not be spinning fast, and it should be possible to apply the rear brake without affecting the engine speed significantly. Conversely, there should be no tendency for the engine to stall due to the idle speed being too low. In rare instances it may not be possible to arrive at a compromise setting, and if this is the case and the machine has covered a lot of miles it may be that wear in the transmission is to blame. For further details refer to the relevant sections of Chapter 1.

4 Cleaning the fuel filter

Although listed as a regular maintenance task, it is only really necessary to clean the fuel filter if water or other contaminants have been introduced into the fuel tank, and are causing problems. In such cases the flow of fuel to the carburettor may be restricted, or water droplets may have passed through the filter and blocked the carburettor passages and jets.

To gain access to the filter it will be necessary to remove the fuel valve from the underside of the tank. The work must be done outside or in a well ventilated area, and well away from any possible fire source. Have ready a metal container into which the fuel can be drained, then remove the left-hand side panel. Disconnect the fuel pipe from the side stub on the fuel valve, and the vacuum pipe from its underside. Position the drain tray beneath the valve, then unscrew the gland nut

Unscrew air filter cover screws (not casing mounting screws on underside) ...

... and lift the cover away

Foam element can be removed for cleaning and re-oiling

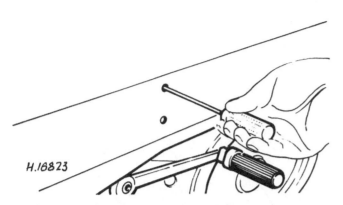

H.16823

Access to idle speed screw – NB50 Vision-X model

which retains it to the underside of the tank. Withdraw the valve and allow the fuel to drain, noting any sediment or water droplets that are expelled with it.

Remove the tubular filter and clean it with compressed air. If contamination was evident, the tank should be flushed through with clean petrol and the contaminated fuel disposed of safely. Refit the fuel valve, taking care not to overtighten the gland nut. Fill the tank with fresh petrol and check that there are no leaks.

Connect the vacuum pipe to its stub, then disconnect it at the inlet adaptor end. Place a container under the fuel outlet stub on the valve and apply suction to the vacuum pipe. In the absence of a vacuum pump this can be done by sucking on the end of the pipe. This should open the valve and allow fuel to flow and will also flush any remaining contaminants from the valve. Bear in mind that if water or dirt has entered the carburettor it may prove necessary to dismantle and clean the carburettor as detailed in Chapter 2.

5 Checking the oil and fuel pipes

With the left-hand side panel removed, examine the oil, fuel and vacuum pipes for signs of damage or deterioration. Remember that fuel leakage is not only wasteful, it can present a fire hazard, whilst leaks in the oil system could lead to lubrication failure and engine damage. If renewal is necessary, always fit pipes of the correct type; if tubing of the wrong material is used it may be attacked by the fuel and oil. Some plastics may become brittle if exposed to fuel, whilst natural rubber tubing will be partially dissolved and the resulting sticky residue will block the carburettor jets and passages. Play safe and obtain the new tubing from a Honda dealer. Note that if the oil pipes are renewed or disturbed they must be bled to remove air as detailed below, or engine damage may result due to lack of lubrication.

6 Checking and bleeding the engine lubrication system

The engine lubrication system does not normally require regular maintenance but should be checked if there is reason to suspect its operation, if the oil tank has been allowed to run dry, or the oil pipes have been disturbed. There is also a filter in the oil tank outlet. Again, this should not normally require attention, but if there is any indication of contaminants in the tank it should be removed, flushed and cleaned.

Check that the pipe between the oil tank and pump is free from air bubbles. If air is present, disconnect the pipe at the pump end and allow the oil to flow through until all air has been displaced. Connect the pipe to the pump inlet stub, then disconnect the delivery pipe at the inlet adaptor. Start the engine and check that oil is forced out of the delivery pipe. On no account run the engine above idle speed and do not leave it running without lubrication for more than one minute.

7 Checking the wheels

The front and rear wheels should be checked periodically and whenever there is reason to suspect damage. Arrange the machine so that the wheel to be checked is clear of the ground, then make up a wire pointer and fix it to the front fork or the transmission casing, as appropriate. Set the pointer to about 1 mm from the wheel rim, then spin the wheel and note any runout. The limit for both axial (side-to-side) and radial (up and down) movement is 2.0 mm (0.08 in). If either wheel exceeds this limit it should be renewed; there is no reliable way to repair it, also the rim or spokes may have become stressed.

Feel for free play by attempting to rock the wheel from side to side. Play in the bearings will be easily felt and indicates the need for renewal. In the case of the front wheel, bearing renewal is described in Chapter 5, whilst the rear wheel is carried on a stub axle housed in the transmission casing, and is covered in Chapter 1.

8 Brake adjustment and lining check

The front and rear brakes must be adjusted periodically to compensate for the gradual wear on the linings. Each brake lever should be able to travel 10 – 20 mm (3/8 – 3/4 in) before brake operation commences. If the free play exceeds this figure, adjust the brake cable at the wheel end, using the adjuster nut on the cable end. Turn it clockwise to reduce free play.

After adjustment, have an assistant apply each brake in turn and check the brake wear indicators at the wheel end. The arrow on the wear indicator arm must not align with the fixed arrow mark; if it does, the brake shoes should be renewed. See Chapter 5 for details. Also check the operation of the starter interlock system: with the rear brake lever locked on, check that the stop lamp bulb is illuminated and that the starter operates; with the rear brake off the starter should not operate.

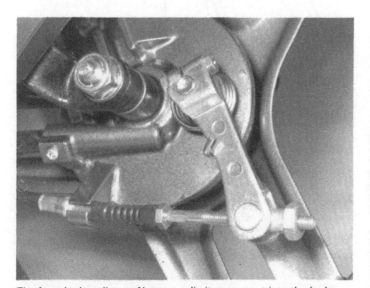

The front brake adjuster. Note wear limit arrow cast into the brake backplate

The rear brake adjuster is located at rear of transmission casing

9   Checking the suspension

Most suspension problems will become evident during riding and the normal pre-ride checks, but it is good practice to give the various suspension parts a closer examination from time to time. As a general rule, any area where free play can be detected will warrant attention. Check for loose fasteners or obviously worn bushes. For more detailed information on suspension checks see Chapter 4.

10   Spark plug renewal

The spark plug should be renewed at the above interval to preclude failure or starting problems. Although it is possible to clean and refit a plug in good condition, it will begin to fail internally in time, and the low cost of a new plug makes this a false economy.

The plug is screwed into the top of the cylinder head and can be reached from the right-hand side of the machine. In the case of the Vision-X model, an access panel is provided inside the luggage compartment, in the right-hand side panel. Disconnect the plug cap and unscrew the plug, using a proper plug spanner. Note that the condition of the old plug can give a good indication of the general condition of the engine – see the colour chart in Chapter 3 for details. Before fitting the new plug, set the electrode gap by bending the outer earth electrode only; never the centre electrode. The gap, which can be checked using feeler gauges, should be 0.6 – 0.7 mm (0.024 – 0.028 in).

Apply a thin film of molybdenum or PBC grease to the plug threads and screw it home finger-tight. Using the plug spanner, tighten it a further $1/4$ to $1/2$ turn only to seal it; beware of overtightening. The recommended plug types are as shown below.

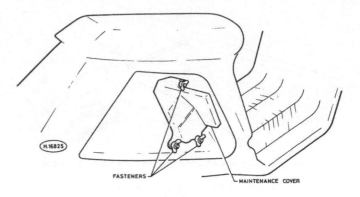

Luggage compartment access panel for spark plug removal – NB50 Vision-X model

	NGK	ND
Standard	BPR6HSA	W20FPR-L
	BPR6HS	W20FPR
Cold climate type	BPR4HSA	W14FPR-L
	BPR4HS	
Extended high-speed type	BPR8HSA	W24FPR-L
	BPR8HS	W24FPR

Yearly, or every 2000 miles (3000 km)

Carry out the operations listed under the 6 monthly/1000 mile heading, then complete the following additional items:

1   Decarbonising the engine and exhaust system

Although listed as a regular maintenance task it is quite likely that the engine and exhaust system will continue to run satisfactorily well beyond the specified mileage, modern two-stroke oils causing far less carbon build-up than earlier types. If the performance of the machine remains normal there is something to be said for leaving well alone, although poor performance or starting problems can often indicate that carbon build-up is becoming significant. To a large extent the rate of carbonisation in the combustion chamber is dependent on the way the machine is used; numerous short journeys will tend to encourage carbonisation, whilst longer trips and the associated higher engine temperatures will be inclined to slow the process.

If you decide to carry out the decarbonisation operation, have ready a new cylinder head gasket.

For normal decarbonising work, it will only be necessary to remove the cylinder head and exhaust system, together with their associated component parts, whilst leaving the cylinder barrel undisturbed. Full details for doing this are contained in the relevant Chapters 1 and 2.

With the piston at the top of the cylinder barrel bore, smear some grease between the edge of the piston crown and the bore surface; this will prevent any carbon deposits from dropping into the bore and make the final cleaning of carbon easier, owing to the fact that the carbon deposits will become stuck to the grease.

Using a hardwood or plastic scraper, remove the carbon deposits from the piston crown. It is desirable to obtain as smooth a finish as is possible, although this is difficult without removing the piston. Take care not to damage the bore surface or piston crown, and avoid the use of sharp instruments, such as screwdrivers, for this reason. The piston crown may, finally, be polished to lessen the likelihood of future carbon build-up and to remove any small remaining traces of carbon;

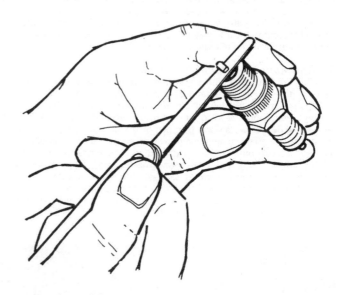

Measuring the spark plug gap

metal polish and a buffing wheel or a piece of rag may be used for this purpose. Ensure all traces of polish are removed prior to reassembly to prevent premature wear of the piston and bore.

The cylinder head may be dealt with in a similar manner. Any stubborn traces of carbon may be removed by using a fine abrasive paper, although great care must be taken not to remove any great amount of metal from the surface thus altering the shape of the combustion chamber.

The exhaust port must also be cleaned of carbon. A build-up of carbon in this area will restrict the flow of exhaust gases from the cylinder.

Before refitting the cylinder head, take care to remove all traces of contamination by washing each component thoroughly with fuel whilst taking care to observe the necessary fire precautions.

Due to the design of the exhaust system and in particular the complex internal substrate of the silencer box, cleaning by conventional means (dissolving deposits with chemicals) is not possible. It is recommended that all carbon deposits be removed from the front end of the exhaust pipe using a suitable scraper and if this fails to improve a blocked system, that the system be renewed.

In skilled hands, it is possible to split open the exhaust system and burn off any deposits with a welding torch, and then weld up the system. Note, however, that this will have the disadvantage of ruining the paint finish and must only be carried out by someone experienced in this type of work. If the system is severely blocked, it is advised that the advice of a Honda dealer is sought.

2 Checking for transmission wear

Initial assessment of transmission wear is best carried out by riding the machine. A developing fault or wear will normally become noticeable during everyday use of the machine. The main symptoms of transmission wear are listed below.

Symptom	Likely cause
1) Transmission inoperative or erratic in operation	a) Worn or slipping drive belt b) Broken or damaged ramp plate in drive pulley c) Broken or weak driven pulley spring d) Worn, contaminated or damaged centrifugal clutch shoe linings e) Damaged splines on driven pulley shaft f) Transmission seized or jammed
2) Transmission creep or frequent stalling of engine	a) Broken or stretched centrifugal clutch shoe spring b) Clutch shoes sticking or seized c) Clutch shoe linings separated from shoe, jamming clutch
3) Machine sluggish when moving off, poor acceleration but high engine speed	a) Worn or contaminated drive belt slipping on pulleys b) Drive pulley rollers worn or sticking c) Drive or driven pulley faces seized or sticking d) Weak or damaged driven pulley spring e) Worn or damaged driven pulley bush f) Clutch shoes worn, contaminated or glazed
4) Poor high speed performance	a) Worn or slipping drive belt b) Worn or damaged drive pulley rollers c) Worn driven pulley bush
5) Unusual transmission noises, burning smell from transmission case	a) Oil or grease contamination of drive belt or pulleys b) Worn drive belt c) Weak driven pulley spring d) Worn or seized driven pulley bush e) Worn clutch shoe pivots

If symptoms such as those described above are noted during riding it will be necessary to remove the transmission cover and investigate the cause of the fault. This procedure is described in detail in Chapter 1, together with information on wear assessment and measurement. If you feel uncertain about tackling these checks, or cannot identify the precise nature of a transmission fault, it is suggested you seek the advice of a Honda dealer before attempting any dismantling work. He should be able to clarify the problem and may suggest the best approach in rectifying it.

3 Checking the steering head bearings

If the steering head bearings become worn or are in need of adjustment it is likely that the steering will feel vague when the machine is ridden. To check this, place the machine on its centre stand and raise the front wheel by placing wooden blocks under the frame spine, below the footboard.

Turn the handlebar from lock to lock. The steering should move smoothly and evenly, with no sign of roughness. If the movement seems 'gritty' or uneven, dismantle the steering column and examine the bearings and races for wear or damage as described in Chapter 4.

Push and pull the handlebar to feel for free play in the steering head bearings. If any clearance can be detected, the bearings are in need of adjustment. This is not a difficult task, but requires a certain amount of dismantling work to gain access to the adjuster. To avoid repetition, reference should be made to Chapter 4.

4 Checking fasteners, wiring and control cables

While the various body panels are removed for other maintenance tasks, give the machine a close check for loose nuts and bolts, tightening them as required to the appropriate torque setting. These are given in the specifications at the beginning of each Chapter. It is a good idea to check other items such as wiring and wiring connectors and cable condition and routing at the same time.

The various control cables should be checked for fraying or wear and renewed as required. Lubricate all cables to maintain smooth and reliable operation. Lubrication can be carried out by detaching the top of the cable and forcing oil through the cable outer. A traditional method is to construct a makeshift funnel by taping a plastic bag around the cable, filling it with oil and allowing it to drain through the cable, preferably overnight. (See the accompanying line drawing.)

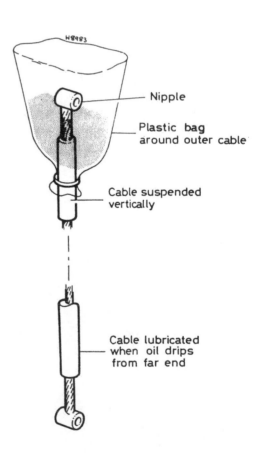

Nipple

Plastic bag around outer cable

Cable suspended vertically

Cable lubricated when oil drips from far end

Oiling a control cable

An alternative is to use a proprietary cable oiler. This can either be one of the hydraulic types where the tool is filled with oil and a screw tightened to force the oil along the cable. Newer versions clamp over the cable and accept the extension nozzle of most maintenance aerosol sprays. Most motorcycle dealers will stock one type or another, but if difficulty is experienced in finding a supplier, try a dealer specialising in off-road and competition machines.

Additional maintenance items

1 Cleaning the machine

Keeping the motorcycle clean should be considered an important part of the routine maintenance, to be carried out whenever the need arises. A machine cleaned regularly will not only succumb less speedily to the inevitable corrosion of external surfaces, and hence maintain its market value, but will be far more approachable when the time comes for maintenance or service work. Furthermore, loose or failing components are more readily spotted when not partially obscured by a mantle of road grime and oil.

The body panels are of moulded plastic construction and need to be treated in a different manner from any metal cycle parts when it comes to cleaning. They will be adversely affected by traditional cleaning and polishing techniques, and lead as a result, to the surface finish deteriorating. Avoid the use of strong detergents which contain bleaching additives, scouring powders or other abrasive cleaning agents, including all but the finest aerosol polishes. Cleaning agents with an abrasive additive will score the surface of the panels thereby making them more receptive to dirt and permanently damaging the surface finish. The most satisfactory method of cleaning the body panels is to 'float' off any dirt from their surface by washing them thoroughly with a mild solution of soapy water and then wiping them dry with a clean chamois leather. A light coat of polish may then be applied to each panel as necessary if it is thought that the panel is beginning to lose its original shine.

The plated parts of the machine should require only a wipe with a damp rag. If they are badly corroded, as may occur during the winter months, when the roads are salted, it is preferable to use one of the proprietary chrome cleaners. These often have an oily base, which will help to prevent the corrosion from recurring.

If the engine parts are particularly oily, use a cleaning compound such as 'Gunk' or 'Jizer'. Apply the compound whilst the parts are dry and work it in with a brush so that it has the opportunity to penetrate the film of grease and oil. Finish off by washing down liberally with plenty of water, taking care that it does not enter the carburettor or the electrics.

Whenever possible, the machine should be wiped down after it has been used in the wet, so that it is not garaged under damp conditions which will promote rusting. Remember there is little chance of water entering the control cables and causing stiffness of operation if they are lubricated regularly as recommended in the preceding Sections of this Chapter.

Chapter 1 Engine and transmission

Refer to Chapter 7 for information relating to the SA50 Met-in

Contents

General description	1
Operations with the engine/transmission in the frame	2
Operations requiring engine/transmission removal from the frame	3
Removing the engine/transmission unit from the frame	4
Dismantling the engine/transmission unit: preliminaries	5
Dismantling the engine/transmission unit: removing the rear wheel	6
Dismantling the engine/transmission unit: removing the flywheel generator	7
Dismantling the engine/transmission unit: removing the carburettor, inlet adaptor, reed valve, oil pump and starter motor	8
Dismantling the engine/transmission unit: removing the cylinder head, barrel and piston	9
Dismantling the engine/transmission unit: removing the transmission cover and kickstart mechanism	10
Dismantling the engine/transmission unit: removing the transmission drive belt and driven pulley/clutch assembly	11
Dismantling the engine/transmission unit: removing the transmission drive (front) pulley	12
Dismantling the engine/transmission unit: removing the transmission final reduction gears	13
Dismantling the engine/transmission unit: separating the crankcase halves	14
Examination and renovation: general	15
Examination and renovation: crankshaft assembly	16
Examination and renovation: main bearings and oil seals	17
Examination and renovation: cylinder head	18
Examination and renovation: cylinder barrel	19
Examination and renovation: piston and rings	20
Examination and renovation: transmission drive pulley unit	21
Examination and renovation: transmission drive belt	22
Examination and renovation: transmission driven pulley and centrifugal clutch unit	23
Examination and renovation: transmission final reduction gearbox components	24
Examination and renovation: kickstart mechanism	25
Reassembling the engine/transmission unit: general	26
Reassembling the engine/transmission unit: joining the crankcase halves	27
Reassembling the engine/transmsision unit: refitting the final reduction gearbox components	28
Reassembling the engine/transmission unit: refitting the transmission components	29
Reassembling the engine/transmission unit: refitting the kickstart mechanism and transmission cover	30
Reassembling the engine/transmission unit: refitting the piston, cylinder barrel and cylinder head	31
Reassembling the engine/transmission unit: refitting the starter motor and oil pump	32
Reassembling the engine/transmission unit: refitting the reed valve and intake adaptor	33
Reassembling the engine/transmission unit: refitting the flywheel generator and engine cowlings	34
Reassembling the engine/transmission unit: refitting the carburettor	35
Reassembling the engine/transmission unit: refitting the rear wheel and exhaust system	36
Reassembling the engine/transmission unit: checking the engine mounting	37
Refitting the engine/transmission unit into the frame	38
Starting and running the rebuilt engine	39

Specifications

Engine

Type	Fan-cooled, single cylinder, two-stroke
Bore	41.0 mm (1.61 in)
Stroke	37.4 mm (1.47 in)
Capacity	49.3 cc (3.0 cu in)
Compression ratio	6.6:1
Maximum power	4.0 ps @ 6000 rpm (DIN)
Maximum torque	0.48 kg/m @ 5000 rpm
Port timing:	
Inlet opening and closing	Controlled by reed valve
Exhaust opens at	76° BBDC
Exhaust closes at	76° ABDC
Scavenge opens at	57° BBDC
Scavenge closes at	57° ABDC

Cylinder head

Maximum warpage	0.10 mm (0.004 in)

Cylinder barrel

Diameter	41.000 – 41.020 mm (1.6142 – 1.6149 in)
Service limit	41.050 mm (1.6162 in)
Maximum warpage	0.10 mm (0.004 in)

Piston and rings

Piston diameter @ 4 mm from bottom of skirt	40.955 – 40.970 mm (1.6124 – 1.6130 in)
Service limit	40.900 mm (1.6102 in)
Cylinder bore to piston clearance	0.035 – 0.050 mm (0.0013 – 0.0019 in)
Service limit	0.10 mm (0.004 in)
Gudgeon pin bore ID	10.002 – 10.008 mm (0.3938 – 0.3940 in)
Service limit	10.025 mm (0.3947 in)
Gudgeon pin OD	9.994 – 10.000 mm (0.3935 – 0.3937 in)
Service limit	9.980 mm (0.3929 in)
Gudgeon pin to piston clearance	0.002 – 0.014 mm (0.0001 – 0.0006 in)
Service limit	0.030 mm (0.0012 in)
Piston ring end gap – top and 2nd	0.10 – 0.25 mm (0.0040 – 0.0100 in)
Service limit	0.50 mm (0.020 in)

Crankshaft assembly

Connecting rod big-end bearing clearance (max):	
Side clearance	0.5 mm (0.02 in)
Radial clearance	0.04 mm (0.002 in)
Crankshaft runout (max):	
Right-hand mainshaft	0.15 mm (0.006 in)
Left-hand mainshaft	0.10 mm (0.004 in)
Small-end eye ID	14.005 – 14.015 mm (0.5514 – 0.5518 in)
Service limit	14.025 mm (0.5522 in)

Transmission

Type	Automatic variable ratio via split pulley and centrifugal clutch connected by toothed V-belt
Transmission ratio range	2.2:1 to 0.87:1
Final reduction ratio	12.106:1
Drive belt width	15.5 mm (0.61 in)
Service limit	14.0 mm (0.55 in)
Drive (front) pulley:	
Bush ID	20.035 – 20.095 mm (0.7887 – 0.7911 in)
Service limit	20.13 mm (0.7925 in)
Drive pulley boss OD	20.005 – 20.025 mm (0.7876 – 0.7884 in)
Service limit	19.97 mm (0.7862 in)
Roller OD	15.92 – 16.08 mm (0.627 – 0.633 in)
Service limit	15.40 mm (0.606 in)
Centrifugal clutch:	
Drum ID	107.0 – 107.2 mm (4.21 – 4.22 in)
Service limit	107.5 mm (4.23 in)
Lining thickness	4.0 – 4.1 mm (0.157 – 0.161 in)
Service limit	2.0 mm (0.08 in)
Driven (rear) pulley:	
Return spring free length	94.2 mm (3.71 in)
Service limit	88.8 mm (3.50 in)
Pulley boss OD – fixed side	33.965 – 33.985 mm (1.3372 – 1.3380 in)
Service limit	33.940 mm (1.3362 in)
Pulley boss ID – moving side	34.000 – 34.031 mm (1.3386 – 1.3398 in)
Service limit	34.066 mm (1.3412 in)
Final reduction method	Gear, housed in separate compartment
Oil grade	SAE 10W/40 motor oil
Oil capacity	90 cc (3.2 Imp fl oz)

Torque wrench settings

Component	kgf m	lbf ft
Cylinder head bolt	0.9 – 1.2	7 – 9
Flywheel rotor nut	3.5 – 4.0	25 – 29
Drive pulley nut	3.5 – 4.0	25 – 29
Inlet stub bolt	0.8 – 1.2	6 – 9
Carburettor bolt	0.9 – 1.2	7 – 9
Clutch centre nut	3.5 – 4.0	25 – 29
Clutch outer nut	3.5 – 4.0	25 – 29
Drive pulley cover bolts	0.25 – 0.40	2 – 3
Engine mounting bolt	3.5 – 4.5	25 – 33
Rear suspension unit lower mounting bolt	2.0 – 3.0	14 – 22

1 General description

The engine/transmission assembly is housed in a light alloy casting which doubles as the rear suspension member, the unit being attached to the frame via a rubber-mounted pivot bracket at the front and by the single rear suspension unit at the rear. This design ensures that the main part of the frame and the rider are effectively isolated from any engine vibration. The main part of the engine/transmission housing is formed by a single large casting which incorporates the transmission housing and the final reduction gears, and forms one half of the crankcase. Separate smaller castings are used to form the right-hand crankcase half and to close the final reduction gearbox and the transmission case.

The engine itself is a simple single-cylinder two-stroke unit featuring reed valve induction and pump-fed lubrication. The crankshaft is a pressed-up assembly comprising a pair of full flywheels arranged each side of the connecting rod and its needle roller big-end bearing. The left-hand crankshaft end protrudes through the crankcase/transmission case casting and carries the transmission drive pulley assembly, whilst the right-hand end carries the flywheel generator assembly, ignition pickup and the engine cooling fan. The crankshaft is supported at each end by journal ball bearings.

The transmission is fully-automatic in operation, comprising a centrifugal clutch and a variable ratio belt drive based on centrifugally-controlled split pulleys, drive from the rear (driven) pulley being transmitted through the final reduction gears to the rear wheel stub axle.

When the engine is stationary or at idle speed, the centrifugal clutch is disengaged and the transmission reverts to its lowest ratio. As engine speed is increased, the centrifugal clutch is gradually brought into engagement and the machine moves off. As the engine speed rises, roller weights in the drive pulley are flung outwards along inclined ramps, closing the pulley halves and increasing its effective diameter. The driven pulley responds by opening against spring pressure, reducing its diameter. This process occurs gradually and is dependent upon both engine speed and load, the result being a fully-automatic transmission system which obviates the need for fixed gear ratios and manual gearchanging.

Starting is by a small starter motor mounted at the front of the unit and operated from a handlebar button. An interlock switch on the rear brake lever prevents starting unless the brake is on, to prevent unintentional departures. A mechanical kickstart lever is provided as a backup to the electric starter.

The enclosed location of the engine means that conventional air cooling cannot be relied upon. Instead, an engine-driven fan draws air in through a grille on the right-hand end of the crankshaft and forces it through ducting around the cylinder and cylinder head. The ducting also dampens down engine noise to a very low level.

2 Operations with the engine/transmission unit in the frame

The engine/transmission unit is well hidden behind the bodywork and at first sight access appears rather restricted. Despite this, it is quite possible to work on many of the engine/transmission components without having to remove the unit from the frame. Where only one area requires attention this is often the best approach, although it should be noted that engine removal is not difficult or time consuming. It follows that if several jobs are to be carried out it is often preferable to remove the engine/transmission unit to obtain better access and a more comfortable working position, and as a general rule this approach is recommended. The various parts and assemblies listed below can be removed with the engine in position.

- a) Cooling fan and cowlings
- b) Cylinder head
- c) Cylinder barrel
- d) Piston and small-end bearing
- e) Flywheel generator and ignition pickup assembly
- f) Air filter
- g) Carburettor
- h) Reed valve
- i) Kickstart mechanism
- j) Transmission belt
- k) Transmission pulleys
- l) Centrifugal clutch
- m) Exhaust system

3 Operations requiring engine/transmission removal from the frame

If the crankshaft assembly, the crankcases or main bearings require attention, it will be necessary to remove the unit from the frame. In addition, in view of the limited access available with the engine/transmission unit in position, it is advisable to remove it to work on the oil pump and the starter motor. Removal of the engine/transmission unit is relatively easy and will take less than an hour to complete.

4 Removing the engine/transmission unit from the frame

1 Place the machine on its centre stand on level ground, leaving room on both sides and to the rear. It is advantageous, although not essential, to raise the machine slightly to permit more comfortable working, and this can be accomplished by building a suitable platform using blocks and some stout planks. The removal operation can be carried out unaided if necessary, although if an assistant is available this will speed things up somewhat and make the actual removal operation considerably easier.

2 Unlock and open the seat, then remove both side panels as described in Routine Maintenance. The seat tends to fall shut, and it is easier to unbolt it from the tank. Removal of the footboard makes access a little easier, but again is not essential.

3 Lift the battery compartment lid and disconnect the battery leads, negative (−) lead first. This avoids any risk of accidental short circuits when disconnecting the electrical leads. If the machine is to be out of service for some time, remove the battery and give it regular refresher charges as described in Chapter 6. Trace the various wires from the engine/transmission unit to the connectors on the left-hand side of the machine. These are housed inside a protective plastic shroud which can be slid clear of the connectors once it has been freed from the frame clip. Separate the various connectors, noting that the leads are colour-coded to facilitate reconnection.

4 Disconnect the spark plug cap, then trace the plug lead back to the coil housing on the top of the air filter. Open the coil housing cover and lift the coil away, placing it on the top of the frame, but leaving the low tension wiring connected. Remove the two bolts which secure the air filter casing to the top of the transmission casing and lift it away, disconnecting the rubber adaptor at the carburettor mouth.

5 Unscrew the carburettor top and withdraw the throttle valve assembly. Place the cable and throttle valve over the frame and clear of the engine. Disconnect the fuel and vacuum pipes from the fuel valve below the tank. Use pliers to work the retaining clips along the pipes, then prise off the pipes using a small screwdriver. Trace the oil pipe from the tank down to the pump. Before disconnecting it, have ready something to plug the end of the pipe to stop the oil in the tank running out; a bolt or a small piece of bar will suffice. Disconnect and plug the pipe, lodging it clear of the engine.

6 Remove the rear brake cable adjuster and displace the cable from the brake arm. Remove the trunnion from the end of the arm and refit it and the adjuster nut on the cable for safe keeping. Disengage the cable from the guide clip under the transmission casing. Remove the two nuts which retain the exhaust pipe to the cylinder barrel and the two bolts which secure the silencer to the crankcase. Lift the exhaust system clear of the engine/transmission unit and place it to one side. Remove the rear suspension lower mounting bolt, leaving the rear of the unit supported by the rear wheel.

7 The engine/transmission unit is now ready for removal and is held in place by the front mounting/pivot bolt only. Before releasing this, check around the unit to make sure that no wires or cables are in a position which would impede removal. If working alone, place wooden blocks under the crankcase to support it while the mounting bolt is removed. Remove the mounting bolt nut and tap the bolt through, steadying the engine as it comes free. The engine/transmission unit, together with the rear wheel, can now be lifted back and out from the left-hand side of the machine and placed to one side to await further dismantling.

5 Dismantling the engine/transmission unit: preliminaries

1 Before any dismantling work is undertaken, the external surfaces of the unit should be thoroughly cleaned and degreased. This will prevent the contamination of the engine internals, and will also make working a lot easier and cleaner. A high flash-point solvent, such as paraffin (kerosene) can be used, or better still, a proprietary engine degreaser such as Gunk. Use old paintbrushes and toothbrushes to work the solvent into the various recesses of the engine castings. Take care to exclude solvent or water from the electrical components and inlet and exhaust ports. The use of petrol (gasoline) as a cleaning medium should be avoided, because the vapour is explosive and can be toxic if used in a confined space.

2 When clean and dry, arrange the unit on the workbench, leaving a suitable clear area for working. Gather a selection of small containers

and plastic bags so that parts can be grouped together in an easily identifiable manner. Some paper and a pen should be on hand to permit notes to be made and labels attached where necessary. A supply of clean rag is also required.

3 Before commencing work, read through the appropriate section so that some idea of the necessary procedure can be gained. When removing the various engine components, it should be noted that great force is seldom required, unless specified. In many cases, a component's reluctance to be removed is indicative of an incorrect approach or removal method. If in any doubt, re-check with the text.

4 Only one special tool is absolutely essential during engine dismantling; this is a generator rotor extractor, Honda part number 07733-0010000. The universal threaded extractor commonly used on lightweight Japanese machines does not fit in this instance. It is theoretically possible to employ a two-legged puller to draw the rotor off the crankshaft end, but there is some risk of distortion, especially if the rotor proves to be a tight fit on the crankshaft taper. If the correct tool cannot be obtained, the part-dismantled unit should be taken to a Honda dealer for the rotor to be drawn off safely.

6 Dismantling the engine/transmission unit: removing the rear wheel

1 This operation can be carried out with the engine/transmission unit in or out of the frame. If the unit is in position in the frame, lock the rear wheel by applying the rear brake and locking it on. The stub axle nut can then be unscrewed. If the engine has been removed, have an assistant hold the rear wheel securely while the nut is slackened.

2 With the stub axle nut removed, pull the wheel off the stub axle splines and place it to one side. The rear brake shoes are now exposed.

If it is intended to carry out further dismantling work it is a good idea to protect the lining surfaces from oil or grease contamination by applying strips of masking tape to the shoe lining surfaces. Do not forget to remove the tape before the wheel is refitted.

7 Dismantling the engine/transmission unit: removing the flywheel generator

1 This operation can be carried out with the engine/transmission unit in or out of the frame. If the unit is installed, it will first be necessary to remove both side panels (see Routine Maintenance) and to disconnect the generator output leads at the connectors on the left-hand side of the machine, near the footboard.

2 Release the bolts which retain the black plastic fan cowling to the crankcase and lift it away. Remove the cooling fan from the rotor after releasing the two retaining screws. It will be necessary to hold the rotor while the retaining nut is slackened. In the absence of the official tool, the home-made equivalent shown in the accompanying photograph can be used. This was made up from $1/8$ in x $1^1/4$ in mild steel strip with bolts for the pivot and holding pegs. When using the tool make sure that the holding pegs do not protrude into the stator too far, or the windings could be damaged.

3 With the rotor immobilised, slacken and remove the retaining bolt. The rotor can now be drawn off its taper using the appropriate Honda extractor or its equivalent (see Section 5) and placed to one side. If the Woodruff key which locates the rotor is loose, remove it and place it with the rotor for safe keeping.

4 The stator and ignition pickup assemblies can now be removed, if required. Each is secured by two bolts to the crankcase. It should be noted at this stage that the ignition is non-adjustable, and thus there is no need to mark the stator for timing purposes.

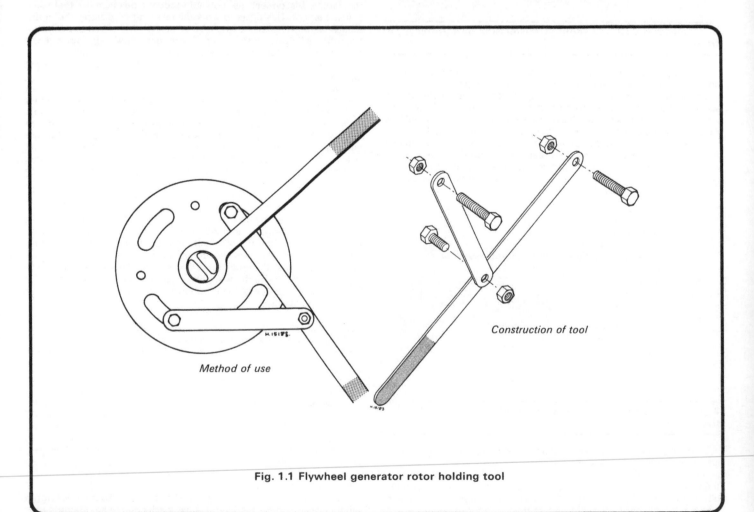

Method of use

Construction of tool

Fig. 1.1 Flywheel generator rotor holding tool

7.2a Release screws and lift away the fan cowling

7.2b The white plastic fan is retained by two screws

7.2c Use home-made holding tool to secure rotor while nut is removed

7.3 Use only the correct extractor to draw the rotor off its taper

7.4 Stator and ignition pickup are each held in place by two bolts

8 Dismantling the engine/transmission unit: removing the carburettor, inlet adaptor, reed valve, oil pump and starter motor

1   The ancillary components listed above should be removed before engine/transmission overhaul. Each item can be removed with the engine/transmission unit in or out of the frame. If the unit is in place, start by detaching both side panels (see Routine Maintenance). Disconnect the fuel and vacuum pipes from the carburettor and the inlet adaptor respectively, by working the retaining clip clear of the stub then prising off the pipes with a small screwdriver. Unscrew the carburettor top and withdraw the throttle valve assembly, lodging it clear of the engine. Trace back the automatic choke wiring to the connectors near the footboard and separate them.

2   Disconnect the oil delivery pipe from the inlet adaptor, plugging the end to exclude dirt and air. Remove the two bolts which secure the carburettor to the inlet adaptor and lift the instrument away.

3   The mounting flange of the inlet adaptor is partially covered by the edge of the cylinder cowling, and it will probably be necessary to release this to allow further progress. Note that one of the cowling bolts also holds a steel retainer which locates the oil pump. Once the cowling is clear of the flange remove the two retaining bolts and lift it away.

4   The reed valve block sits in a recess below the adaptor flange and can now be prised gently up and clear of the crankcase. It is almost inevitable that the reed valve gaskets will be damaged during removal and these should be renewed during installation. Place a piece of clean rag in the inlet port to prevent dirt from entering the engine; do not forget to remove this before the reed valve is refitted.

5   As mentioned above, the oil pump is retained by a plate secured by one of the cylinder cowling bolts. If still in place this should be removed and the retainer plate lifted away. Disconnect and plug the oil feed pipe, where this is still in place. The oil pump is now located by its O-ring only and can be pulled clear of the crankcase.

6   To free the starter motor, trace back the motor leads to the connector near the left-hand edge of the footboard and separate it. Remove the two motor retaining bolts and pull the motor to the right until it comes free of the crankcase.

8.2a Disconnect and plug the oil delivery pipe at the inlet adaptor

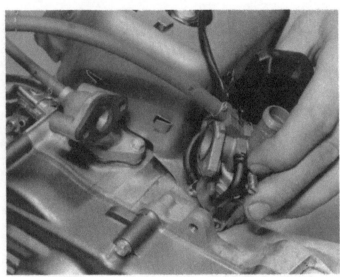

8.2b Remove the two mounting bolts and remove the carburettor

8.3a Note that one of the cowling bolts also holds the oil pump retainer

8.3b With cylinder cowling removed, the adaptor can be unbolted and lifted away

8.4 Lift reed valve out of inlet tract and plug crankcase with clean cloth

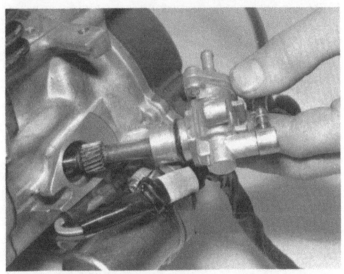

8.5 Once retainer plate has been removed, the oil pump can be withdrawn from crankcase

8.6 The starter motor can now be removed after releasing its two retaining bolts

9 Dismantling the engine/transmission unit: removing the cylinder head, barrel and piston

1 The above components can be removed with the engine/transmission unit in the frame or on the workbench. If the unit is installed in the frame it will first be necessary to remove the following:

a) Both side panels (see Routine Maintenance)
b) The air filter case (see Section 4)
c) The fan grille (see Section 7)
d) The exhaust system (see Section 4)
e) The cylinder cowling (see Section 8)

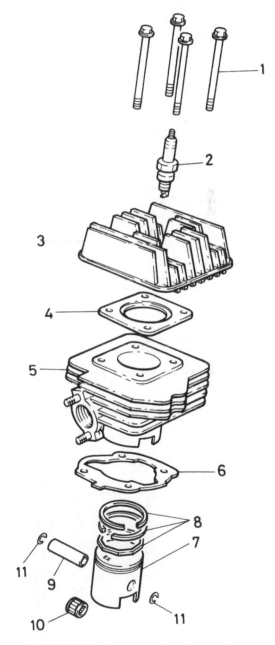

Fig. 1.2 Cylinder head, barrel and piston

1 Bolt – 4 off	7 Piston
2 Spark plug	8 Piston rings
3 Cylinder head	9 Gudgeon pin
4 Cylinder head gasket	10 Small-end bearing
5 Cylinder barrel	11 Circlip – 2 off
6 Base gasket	

2   It is a good idea to loosen the spark plug before the cylinder head is removed. Slacken each of the four cylinder head bolts by $1/4$ turn each in a diagonal sequence until pressure on the head is released. Remove the bolts, then lift the head away. If it is stuck to the barrel, tap around the side of the head with a soft-faced mallet to loosen it. **Do not** try to prise it off with a screwdriver or damage to the gasket face may be caused.

3   Lift the barrel slightly until some clean cloth can be placed in the crankcase mouth. This will prevent any debris entering the crankcase as the barrel is removed. Note that if this precaution is not observed and dirt enters the crankcase it is essential to completely dismantle the unit to remove the dirt. Continue lifting the barrel and support the piston as it emerges from the cylinder bore.

4   Using a pair of snipe-nose pliers, grasp the gudgeon pin circlip via the cutout in the piston and remove it. Alternatively, the circlip can be prised out using a small electrical screwdriver. The remaining circlip can be removed in a similar fashion, if required. It is normally quite easy to displace the gudgeon pin by pushing it out from one side of the piston, but if it proves to be unusually tight, proceed as follows.

5   Obtain a container of boiling or near-boiling water and soak a piece of clean rag in it. Wearing gloves and using pliers to avoid scalding, and taking great care, remove the rag and twist it to squeeze out excess water. Wrap the rag around the piston and hold it there for about one minute. The light alloy piston will expand when heated, loosening its grip on the gudgeon pin which, being steel, expands much less. Remove the rag and press out the pin with a drift. Lift away the piston and displace the needle roller small-end bearing, placing it inside the piston to prevent its loss.

6   Where the need arises, the gudgeon pin can be pressed out using a drawbolt arrangement as shown in the accompanying line drawing. It is unlikely that this method will be necessary unless the gudgeon pin or small-end bearing have suffered damage due to seizure. **On no account** try to drift the pin out; this will almost certainly damage the connecting rod or crankshaft.

10 Dismantling the engine/transmission unit: removing the transmission cover and kickstart mechanism

1   The above components can be removed with the engine/transmission unit in or out of the frame. If the unit is in position, start by removing the left-hand side panel and the air filter casing, and free the rear brake cable from the clip below the transmission casing. **Note:** *On*

no account operate the starter motor once the transmission cover has been removed or damage may result. It is advisable to disconnect the battery leads to make sure that this cannot happen.

2   Using a spirit-based felt marker-pen, mark the position of the two air filter retaining screws in relation to the cover; this will save a good deal of confusion during reassembly. Remove the remaining screws and lift the transmission cover away. If the cover proves to be stuck, tap around the joint with a soft-faced mallet or a hardwood block and hammer to jar it free.

3   Remove the kickstart lever, and free the circlip which retains the shaft in the casing bore. Temporarily refit the lever, then turn it until the kickstart driven gear and friction clip assembly can be removed. Gently release the lever and remove it from the shaft. Working from inside the cover, turn the gear to take up the return spring pressure and pull the assembly part way out of the cover. Once the gear is clear of the stop, allow it to turn until spring tension is released, then displace the assembly and lift it away.

11 Dismantling the engine/transmission unit: removing the transmission drive belt and driven pulley/clutch assembly

1   The drive belt can be removed with the engine/transmission unit in or out of the frame. If the unit is in position, start by removing the transmission cover as described in Section 10.

2   Remove the clutch drum retaining nut from the rear pulley and centrifugal clutch assembly. This will necessitate holding the drum while the nut is slackened, either by using the home-made holding tool previously described for holding the generator rotor, or a chain or strap wrench around the drum periphery. With the drum immobilised, slacken and remove the retaining nut and lift the drum away. Disengage and remove the starter pinion assembly from the front of the transmission casing.

3   The transmission will be under tension from the driven pulley return spring and this must be compressed as follows. Grasp the belt just forward of the rear pulley assembly and squeeze it together. As the belt tightens around the pulley, the pulley halves will be forced apart against spring pressure, leaving slack at the front of the belt. Hold the belt in this position and lift the rear pulley/clutch assembly away from the shaft, then release the belt from the pulleys.

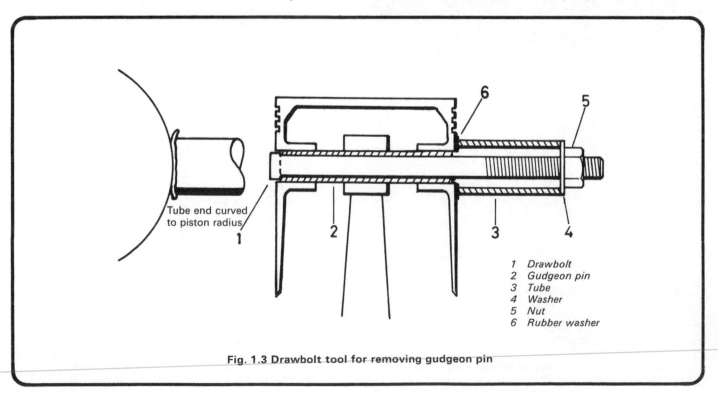

Tube end curved to piston radius

1   Drawbolt
2   Gudgeon pin
3   Tube
4   Washer
5   Nut
6   Rubber washer

Fig. 1.3 Drawbolt tool for removing gudgeon pin

12 Dismantling the engine/transmission unit: removing the transmission drive (front) pulley

1 The drive pulley can be removed with the engine/transmission unit in or out of the frame. If the unit is in position, start by removing the transmission cover as described in Section 10 and the drivebelt as described in Section 11.
2 If the unit is installed in the frame, remove the right-hand side panel, the fan grille and fan, then use the home-made holding tool described previously to lock the crankshaft while the drive pulley nut is slackened.
3 If the drive clutch is being removed as part of a full overhaul, lock the crankshaft by passing a round bar through the connecting rod small-end eye and supporting the ends of the bar on wooden blocks placed against the crankcase mouth (see accompanying photograph). Once the nut has been removed, the fixed outer face of the drive pulley can be lifted away, followed by the inner face and roller assembly.

13 Dismantling the engine/transmission unit: removing the transmission final reduction gears

1 The final reduction gears can be removed with the engine/transmission unit in or out of the frame. If the unit is in position, start by removing the transmission cover as described in Section 10, then remove the drive belt and rear pulley/clutch assembly as detailed in Section 11. Remove the rear wheel.
2 Place a drain tray below the end of the transmission casing and remove the five gear cover bolts. Separate the cover slightly and allow the oil to drain. Lift away the cover and remove the gasket and the two dowel pins. The input shaft will come away with the cover, leaving the idler shaft and output shaft in the casing. Remove the thrust washer from the end of the idler shaft and place it with the cover for safe keeping.
3 Remove the circlip from the end of the output shaft and lift away the large final driven gear. The idler shaft assembly can now be withdrawn from the casing, as can the output shaft. The input shaft and the various bearings need not be disturbed unless inspection shows them to be in need of attention.

14 Dismantling the engine/transmission unit: separating the crankcase halves

1 This operation requires that the engine/transmission unit be removed from the frame and the ancillary parts stripped as described in the preceding Sections. You will also need to make up a simple tool to draw off the crankcase right-hand half without risk of damage, and this can be made up as follows. Obtain a piece of steel plate of the following approximate dimensions: $1/4$ x $1 1/2$ x 4 in (6 mm x 38 mm x 100 mm) and drill two $1/4$ in (6 mm) holes as shown in the accompanying illustration.
2 Lay the unit on the workbench with the crankcase right-hand half uppermost, and remove the six crankcase holding bolts. Place two of the longer crankcase bolts through the holes in the extractor plate and screw them into the threaded holes provided in the crankcase. Tighten the two bolts evenly to draw the crankcase half off the crankshaft.
3 To remove the crankshaft from the remaining crankcase half, invert it and support it on wooden blocks so that the crankshaft end is several inches above the workbench. Place a wad of rag under the crankshaft to prevent it falling onto the hard work-surface. Run the drive clutch nut onto the crankshaft threads to protect them and tap the crankshaft through the bearing.
4 If it proves particularly tight, do not be tempted to hit the crankshaft hard or it may become misaligned. To counter this, place a wooden wedge between the flywheel faces, opposite the crankpin. If this fails, either have the crankshaft pressed out, or make up a second extractor from scrap steep plate and use it to press the crankshaft out evenly.

13.2 Output and idler shafts can be removed after final reduction gearbox cover has been detached

14.2a Home-made tool for drawing off the crankcase right-hand half (see text)

14.2b Right-hand main bearing will probably stay in place on the crankshaft

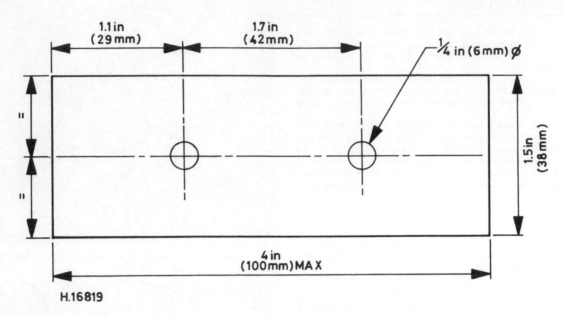

Fig. 1.4 Crankcase right-hand half removal tool dimensions

15 Examination and renovation: general

1   Before examining the parts of the dismantled engine unit for wear it is essential that they should be cleaned thoroughly. Use a petrol/paraffin mix or a high flash-point solvent to remove all traces of old oil and sludge which may have accumulated within the engine. Where petrol is included in the cleaning agent normal fire precautions should be taken and cleaning should be carried out in a well-ventilated place.

2   Examine the crankcase castings for cracks or other signs of damage. If a crack is discovered, it will require a specialist repair.

3   Examine carefully each part to determine the extent of wear, checking with the tolerance figures listed in the Specifications section of this Chapter or in the main text. If there is any doubt about the condition of a particular component, play safe and renew.

4   Use a clean lint-free rag for cleaning and drying the various components. This will obviate the risk of small particles obstructing the internal oilways, and causing the lubrication system to fail.

5   Should any studs or internal threads require repair, now is the appropriate time to attend to them. Where internal threads are stripped or badly worn, it is preferable to use a thread insert, rather than tap oversize. Most dealers can provide a thread reclaiming service by the use of Helicoil threads inserts. They enable the original component to be re-used.

6   Sheared studs or screws can usually be removed with screw extractors, which consist of tapered, left-hand thread screws, of very hard steel. These are inserted by screwing anticlockwise, into a pre-drilled hole in the stud, and usually succeed in dislodging the most stubborn stud or screw. The only alternative to this is spark erosion, but as this is a very limited, specialised facility, it will probably be unavailable to most owners. It is wise, however, to consult a professional engineering firm before condemning an otherwise sound casing. Many of these firms advertise regularly in the motorcycle papers.

16 Examination and renovation: crankshaft assembly

1   The crankshaft assembly comprises two full flywheels, two mainshafts, a crankpin and big-end bearing, a connecting rod and a caged needle roller small-end bearing. The general condition of the big-end bearing may be established with the assembly removed from the engine, or with just the cylinder head and barrel removed. In this way it is possible to decide whether big-end renewal is necessary, without a great deal of exploratory dismantling.

2   Big-end failure is characterised by a pronounced knock which will be most noticeable when the engine is worked hard. The usual causes of failure are normal wear, or a failure of the lubrication supply. In the case of the latter, big-end wear will become apparent very suddenly, and will rapidly worsen. Check for wear with the crankshaft set in the TDC (top dead centre) position, by pushing and pulling the connecting rod. No discernible movement will be evident in an unworn bearing, but care must be taken not to confuse this with endfloat, which is normal, and main bearing wear.

3   If the crankshaft assembly is to be examined with it removed from the engine, then the degree of wear in the big-end can be assessed with a greater amount of accuracy if the assembly is first washed in fuel to remove any residual oil.

4   The big-end bearing is, like the small-end bearing, of the caged needle roller type and under normal circumstances will give good service. If a dial gauge is readily available, an accurate measurement of radial wear in the big-end bearing may be made by positioning the gauge pointer at the positions indicated in the figure accompanying this text. If the measurement taken exceeds the service limit of 0.04 mm (0.002 in), then the bearing must be renewed.

5   Measure the big-end side clearance by inserting the blade of a feeler gauge between the side of the connecting rod at the big-end eye and the crankshaft flywheel. If the clearance measured exceeds the service limit of 0.50 mm (0.02 in), then the bearing must be renewed.

6   Measurement of crankshaft deflection should only be made with the crankshaft assembly removed from the crankcase and set up on V-blocks which themselves have been positioned on a completely flat surface. The amount of deflection present in each mainshaft should not exceed the service limit of 0.10 mm (0.004 in) in the left-hand shaft or 0.15 mm (0.006 in) in the right-hand shaft. Each measurement should be made at the point on the shaft indicated in the figure accompanying this text. If either one of these service limits is exceeded, then the complete crankshaft assembly must be replaced with a serviceable item.

7   Like the big-end bearing, the small-end bearing should seldom give trouble unless a lubrication failure has occurred. Wash the small-end bearing thoroughly in petrol and then push it into the small-end eye of the connecting rod. Push the gudgeon pin through the bearing. Hold the connecting rod steady and feel for any discernible movement between it and the gudgeon pin. If movement is detected, do not automatically assume that the bearing is worn but

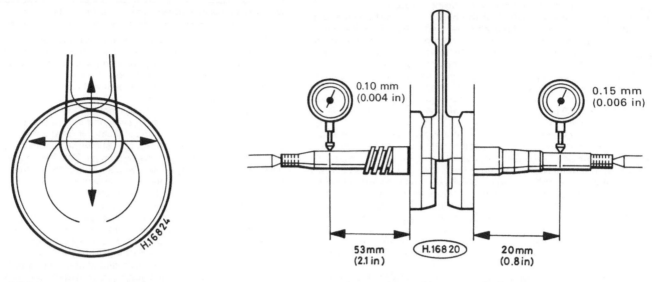

Fig. 1.5 Measuring big-end bearing radial wear

Fig. 1.6 Measuring crankshaft runout

check that the bore of the small-end eye and the outer diameter of the gudgeon pin are not worn beyond the service limits given in the Specifications Section of this Chapter. If both the connecting rod and gudgeon pin are found to be within limits, discard the bearing and replace it with a new item. Close inspection of the bearing will show if the roller cage is beginning to crack or wear, in which case the bearing must be renewed, even though play may be within the given limits.

8 If any fault is found or suspected in any one of the component parts which make up the crankshaft assembly, then it is recommended that the complete crankshaft assembly is taken to a Honda dealer, who will be able to confirm the worst and supply a new or service-exchange assembly. The big-end bearing is reached by dismantling the pressed-up flywheel assembly and this is a job for a specialist. On completion of bearing renewal, the assembly must be trued up and balanced, a job requiring skills and facilities beyond the reach of most home mechanics. Much of the cost of reconditioning the big-end assembly will have been saved by removing the engine and separating the crankcases.

9 Finally, do not omit to check the condition of the worm gear which forms part of the right-hand mainshaft. If this gear is seen to be broken or badly worn, then the right-hand mainshaft will have to be renewed.

17 Examination and renovation: main bearings and oil seals

1 The crankshaft main bearings should be washed out with clean petrol to remove all traces of oil and grease, and then checked for free play. Each bearing can be checked in position. A small amount of axial and radial movement is normal and acceptable, but excessive play or roughness when the bearing is turned indicates the need for renewal. It is advisable not to attempt to remove either of the bearings unless they are to be renewed.

2 The left-hand main bearing will probably have remained in the crankcase boss during crankcase separation, whilst the right-hand bearing will stay in place on the crankshaft end. The left-hand bearing and seal can be removed by placing the crankcase on the workbench, supported on wooden blocks around the main bearing boss. In the absence of the official driving tool and attachments, use a length of tubing as a drift to tap the bearing and seal out of the casing. The bearing should come out quite easily, but if it should prove tight, try warming the bearing boss area in near-boiling water to expand the alloy before driving the bearing out. The new bearing should be driven home squarely using a large socket as a drift. Note that the drift should bear only on the bearing outer race during installation or damage may result. Fit a new oil seal in a similar fashion, having first greased the sealing lip.

3 To remove the right-hand bearing a suitable extractor will be required. These can often be hired from tool hire shops, or it may be possible to borrow one from a friendly motorcycle dealer. Failing this, take the crankshaft to a Honda dealer for the old bearing to be drawn off. The new bearing should be fitted into the right-hand crankcase half as described above.

18 Examination and renovation: cylinder head

1 Check that the fins of the cylinder head are not clogged with oil or dirt. If heavy contamination is present, then it is likely that the engine will overheat. If necessary, use a degreasing agent and brush to clean between the fins. Check that no cracks are evident, especially in the vicinity of the holes through which the holding down bolts pass, and near the spark plug threads.

2 Check the condition of the thread in the spark plug hole. If it is damaged an effective repair can be made using a Helicoil thread insert. This service is available from most Honda dealers. The cause of a damaged thread can usually be traced to overtightening of the plug or using a plug of too long a reach. Always use the correct plug and do not overtighten.

3 If leakage problems have been experienced between the cylinder head and cylinder barrel mating surfaces, the cylinder head should be checked for distortion by placing a straight-edge across several points on its mating surface and then attempting to slide a 0.10 mm (0.004 in) feeler gauge between the straight-edge and the mating surface.

4 If the cylinder head is found to be warped beyond this limit, grind it flat by placing a sheet of emery paper on a surface plate or sheet of plate glass and rubbing the cylinder head mating surface against it, in a slow, circular motion. Commence with 200 grade paper and finish with 400 grade paper and oil. If it is found necessary to remove a substantial amount of metal before the mating surfaces become completely flat, obtain advice from a Honda dealer as to whether it is necessary to obtain a new head.

5 Note that most cases of cylinder head distortion can be traced to unequal tensioning of the cylinder head securing bolts and to tightening the bolts in the incorrect sequence.

19 Examination and renovation: cylinder barrel

1 The usual indication of a badly worn cylinder barrel and piston is known as piston slap, a metallic rattle that occurs when there is little or no load on the engine.

2 Commence by cleaning the outside of the cylinder barrel, taking care to remove any accumulation of dirt from between the cooling fins. Carefully remove the ring of carbon from the mouth of the bore, so that

an accurate assessment of bore wear can be made.

3   A close visual examination of the bore surface must be made, to check for scoring or any other damage, particularly if broken piston rings were encountered during the stripdown. Any damage of this nature will necessitate renewal of the barrel, as it is impossible to obtain a satisfactory seal if the bore is not perfectly finished.

4   There will probably be a lip at the uppermost end of the cylinder bore which marks the limit of travel of the top piston ring. The depth of the lip will give some indication of the amount of bore wear that has taken place even though the amount of wear is not evenly distributed.

5   The most accurate method of measuring bore wear is by the use of a cylinder bore DTI (dial test indicator) or a bore micrometer. Measurements should be made at the top of the bore (just below the wear lip), at a point in the bore just above the ports and at a point just above the cutout in the bore spigot, Make measurements in both a fore and aft and a side to side plane at these positions. If the largest measurement taken exceeds the service limit of 41.050 mm (1.6162 in), then the barrel must be renewed as no rebore sizes are given and no oversize pistons or rings are available. Note that the barrel will have the letter A or B stamped on its crankcase mating surface. The replacement barrel must carry the same letter as the item it is to replace.

6   It is however, most unlikely that the average owner will have the above mentioned items of equipment readily available. A slightly less accurate but more practical method of measurement is as follows. It is possible to determine the amount of wear in the bore by inserting the piston, devoid of rings, into the bore, so that its crown is positioned just below the wear lip at the top of the bore. Measure the distance between the wall of the bore and the top side of the piston by using feeler gauges.

7   Move the piston down the bore so that its crown is positioned in line with a point just above the cutout in the bore spigot. Repeat the measurement. Doing this, and subtracting the lesser measurement from the greater, will give the difference between the bore diameter in an area where the greatest amount of wear is likely to occur and an area in which there should be little or no wear. If the difference found exceeds 0.050 mm (0.002 in), then the barrel will have to be renewed. Bear in mind that the curvature of the gap being measured will tend to preclude accurate measurement by this method but it should be possible to gain a good indication of whether renewal of the barrel is necessary. In view of the cost involved in purchasing a new barrel, it is well worth having one's worst suspicions confirmed by asking a competent engineer to make an accurate measurement of the bore diameter.

8   Upon receiving a new cylinder barrel, check to confirm that the edges of the ports, where they meet the surface of the bore, have been given a slight chamfer. This chamfer is necessary to prevent the possibility of the piston rings catching on the unchamfered edge of a port and breaking, thus necessitating a further strip down and barrel renewal. Chamfering of the port edges can be carried out by very careful use of a metal scraper but it is essential to ensure that the wall of the bore does not become damaged in the process. Finish off any chamfer made by polishing its cut edges with fine emery paper.

9   Establishment of the clearance between the cylinder bore and the piston can be made either by direct measurement of the cylinder bore and piston diameters, and then by subtracting the latter figure from the former, or by actual measurement of the gap using a feeler gauge. In either case, if the clearance exceeds the maximum wear limit of 0.10 mm (0.004 in), then there is evidence that either a new piston or a new cylinder barrel is required.

10 If the method of direct measurement of the piston and cylinder bore is decided upon, then the measurement of piston diameter should be made at a point 4 mm (0.16 in) from the base of the piston skirt, at right-angles to the gudgeon pin hole. Full details of the method of measuring bore wear are given in paragraph 5 of this Section.

11 Finally, check the cylinder barrel to cylinder head mating surface for distortion by placing a straight-edge across several places on it and attempting to slide a 0.10 mm (0.004 in) feeler gauge between the straight-edge and mating surface. If the cylinder barrel proves to be warped beyond this limit, grind it flat by placing a sheet of emery paper on a surface plate or sheet of plate glass and rubbing the mating surface against it, in a slow, circular motion. Commence this operation with 200 grade paper and finish with 400 grade paper and oil. If it is thought necessary to remove a substantial amount of metal in order to bring the mating surface back to within limits, obtain advice from a Honda dealer as to whether it is necessary to obtain a replacement barrel.

20 Examination and renovation: piston and rings

1   If inspection of the cylinder barrel has revealed damage, or wear beyond the specified service limit, it is unlikely that the existing piston will be serviceable, but if bore condition is acceptable, check the piston and rings as described below. It goes without saying that if the piston is obviously badly worn or damaged it will have to be renewed, even if the cylinder bore is serviceable.

2   Start by removing the piston rings. This can be done by very carefully spreading the ends of each ring in turn and easing it off the piston. Whilst this can be done manually, it is safer and often easier to arrange several thin metal strips as shown in the accompanying line drawing and sliding the rings over them. This method is particularly useful where the rings have become gummed in their grooves by accumulated carbon. Old feeler gauge blades are ideal for this purpose.

3   As each ring is removed, note the etched markings near the gap and on the upper edge. On a well-used machine they may be very faint, so as a precaution mark each ring with a spirit-based felt pen to indicate the groove to which it belongs and the top face. The rings, if they are to be refitted, must be installed in their original grooves.

4   Over a period of time the piston will wear, this being most obvious near the bottom of the piston skirt and at right-angles to the gudgeon pin. The appearance of the piston surface will vary between lightly polished to seriously worn or scored, according to the age of the machine and its past history. The extent of wear can be measured using a micrometer, the measurement being taken at a point 4 mm from the bottom of the skirt and at 90° to the gudgeon pin bore. If the piston has worn to 40.900 mm (1.6102 in) or less, it must be renewed.

5   Where partial seizure has occurred, the piston surface may have 'picked up' slightly. Provided that the damage is not too severe and that the piston is otherwise serviceable, the roughness may be dressed off using a fine file or abrasive paper. Take care not to remove any more material than is necessary to achieve a smooth surface.

6   Clean out any accumulated carbon from the piston ring grooves using a suitable scraper. A section of broken piston ring is ideal for this job, but failing that a small electrical screwdriver can be used. Take great care not to enlarge the grooves during cleaning. Honda do not provide a figure for the ring to groove clearance, but if there is excessive play, ring breakage, loss of compression or engine noise may result. As a guide, there should be barely perceptible free play between the ring and its groove, and if this exceeds about 0.1 mm (0.004 in) with an unworn ring it may be necessary to renew the piston. If in doubt, seek the advice of a Honda dealer.

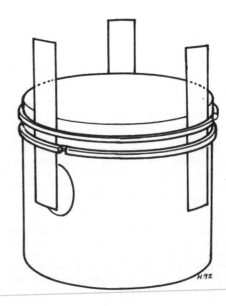

Fig. 1.7 Method of removing gummed piston rings

7 Check that the small pegs which locate the ends of the piston rings are secure and undamaged. The pegs are of vital importance and if loose could allow the ring to rotate on the piston. If this happens, the ring ends will get trapped in one of the ports and the ring will break. If necessary, renew the piston to avoid more serious damage occurring later.

8 The rings themselves should present an even, shiny surface on their working faces, without any signs of scores or discoloration. Any discoloured patches indicate leakage of gas past the rings and thus the need for renewal. Ring wear is assessed by placing each ring in turn into the bottom of the cylinder bore and measuring the ring end gap. Use the base of the piston to ensure that the ring sits square in the bore, and position it 30 mm (1¼ in) from the bottom of the bore. If the end gap of either exceeds 0.50 mm (0.020 in) renew the rings as a pair.

9 If new rings are to be fitted into a used bore, it is important to note that there is a risk of ring breakage if the top ring contacts the wear ridge. This can be removed using a de-ridging tool, or by honing, and this should be left to a specialist. It will also be necessary to remove the smooth, glazed surface of the bore to enable the new rings to bed in, and this 'glaze busting' operation should be carried out together with any de-ridging required. This work can be carried out by any competent engineering firm, and most Honda dealers will be able to arrange this service if required. Do bear in mind that if the honing work is likely to take the bore past its service limit it will be necessary to renew the barrel and piston anyway, so consider this before having the job done.

10 Examine the gudgeon pin and its bore for signs of wear. Service limit figures for the outside diameter of the pin, inside diameter of the gudgeon pin bore, and the clearance between the two are given in the Specifications section. To check these measurements a micrometer and a bore micrometer will be required. Few owners will possess this equipment, but it will suffice to check that there is little or no discernible free play between the piston and gudgeon pin. If movement can be felt, consult a Honda dealer for advice.

11 Fit the small-end bearing and the gudgeon pin into the small-end eye and feel for clearance between them. If any movement is noted, renew the bearing and also the gudgeon pin if this has been marked by the bearing rollers.

21 Examination and renovation: transmission drive pulley unit

1 Place the sliding face assembly of the drive pulley unit on a clean work surface and remove the three bolts which retain the cover plate in position. Removal of this plate will reveal the ramp plate located beneath it. Ease this plate out of the face housing and detach each one of the three nylon slide pieces from its outer edge. Place the complete assembly to one side, ready for inspection.

2 Lift each one of the six rollers out of the face housing and wipe them clean before placing them to one side, ready for inspection. Wipe any grease out of the housing and detach the O-ring from its outer edge. Push the inner sleeve out of the centre of the housing.

3 The face housing should now be closely examined for signs of fatigue or failure, such as presence of hairline cracks around the base of the webs within the housing or around the threaded bosses of the cover plate retaining bolt locations.

4 Degrease the surface of the bush fitted into the centre of the face housing and degrease the outer surface of the drive face boss. Refit the boss into the bush and check for movement between the two components. If excessive movement is found, one or both of the components will have to be renewed. Measure the overall diameter of the boss. If the measurement taken is less than the service limit of 19.97 mm (0.786 in) then the boss must be renewed. The internal diameter of the bush should not be more than the service limit of 20.130 mm (0.7925 in). If the bush is worn beyond this limit then it must be renewed. Examine the condition of the O-ring detached from the housing. If the ring is flattened or in any way damaged, then it must be renewed. Fit the serviceable O-ring to the housing.

5 Carefully examine the bearing surface of each roller sleeve. If any one sleeve is scored or shows signs of having developed a flat spot, then all six sleeves should be renewed. The same applies if the overall diameter of any one sleeve is found to be less than the service limit of 15.40 mm (0.606 in) or if a sleeve is found to be anything other than a firm fit on its roller.

6 Examine the nylon slide pieces removed from the ramp plate for signs of damage or wear. Each piece should be a firm fit on the ramp plate and a close sliding fit over its retaining web in the face housing. If any one piece is found to be defective, then all three pieces should be renewed. Examine the ramp plate for signs of fatigue or failure before fitting the serviceable slide pieces to its outer edge. Finally, examine the cover plate for signs of fatigue.

7 Reassemble the drive pulley unit by reversing the dismantling procedure. Lubricate each roller ramp before inserting the rollers into the face housing. With the rollers all fitted, smear the surface of the centre bush and the inside surface of the ramp plate with grease before pushing the plate into position. The grease used should be of a lithium-based type. Check, before fitting and tightening the cover plate retaining bolts, that the plate is pushed fully down over the O-ring and that the O-ring remains in its location. These bolts should be tightened to a torque loading of 2 – 3 lbf ft (0.25 – 0.40 kgf m).

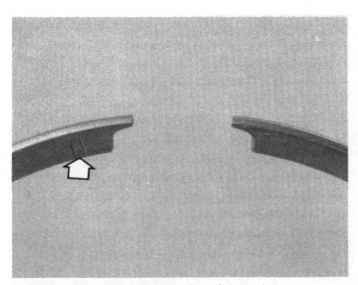

20.3 Note 'N' marking on the upper face of piston ring

21.1a Remove the three bolts and lift away the top cover ...

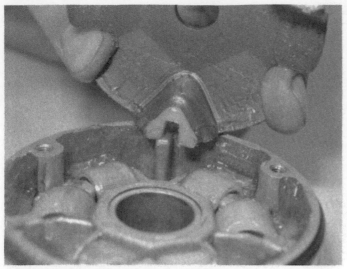

21.1b ... to allow ramp plate to be removed

21.2 The rollers can now be removed and cleaned

21.5 Check each roller for wear or damage

21.7a Fit a new O-ring if worn or broken, and grease well before fitting top cover

21.7b Grease and install the inner sleeve

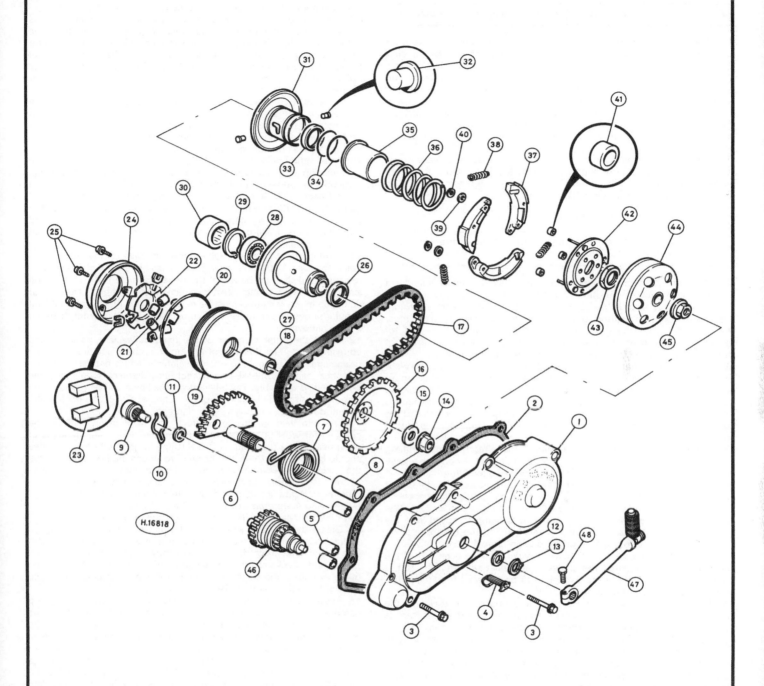

Fig. 1.8 Transmission clutch and pulley assemblies

1	Transmission cover	13	Circlip	25	Bolt – 3 off	37	Clutch shoe – 3 off
2	Gasket	14	Nut	26	Oil seal	38	Spring – 3 off
3	Bolt – 8 off	15	Washer	27	Rear pulley boss	39	Washer – 3 off
4	Cable guide	16	Front pulley outer half	28	Outer bearing	40	Circlip – 3 off
5	Dowel pin – 2 off	17	Drive belt	29	Circlip	41	Damping rubber – 3 off
6	Kickstart shaft	18	Inner sleeve	30	Inner bearing	42	Clutch shoe plate
7	Kickstart spring	19	Front pulley inner half	31	Rear outer pulley half	43	Nut
8	Sleeve	20	O-ring	32	Guide pin – 2 off	44	Clutch drum
9	Kickstart driven gear	21	Roller – 6 off	33	Oil seal	45	Nut
10	Friction clip	22	Ramp plate	34	O-ring – 2 off	46	Starter pinion
11	Thrust washer	23	Nylon slide piece – 3 off	35	Seal collar	47	Kickstart lever
12	Washer	24	Top cover	36	Pulley spring	48	Pinch bolt

22 Examination and renovation: transmission drive belt

1 The drive belt should be examined for signs of wear or deterioration, such as fraying or delamination of the belt driving surface from the underlying reinforcing layers. If such signs are evident, the belt should be renewed. The components in the transmission casing should run dry, so look for signs of oil or grease contamination on the belt or pulleys. If contaminated, the belt may slip under acceleration and will soon start to wear. Early indications are unusual noises or smells from the transmission area. Make sure that the cause of the contamination is located and remedied before a new belt is fitted.

2 In time, the rubbing surfaces of the belt will begin to wear down and will eventually cause slipping. To check this, measure across the widest point on the belt. A new belt measures 15.5 mm (0.61 in) in width, the service limit being 14.0 mm (0.55 in).

23 Examination and renovation: transmission driven pulley and centrifugal clutch unit

1 The centrifugal clutch and the driven pulley effectively comprise a single unit which is mounted on the input shaft to the reduction gearbox. If it is found that abnormally high engine speed is required before power is transmitted to the rear wheel, then the clutch unit must be removed from the machine and the shoe linings inspected for excessive wear, scoring, or contamination by oil or grease. If any of these factors are evident, then the shoes will require renewal.

2 If it is found that the machine has a tendency to creep forward when the engine is running at tickover speed, the clutch shoe return springs may have weakened or stretched, thus allowing the shoes to come into engagement prematurely with the clutch drum. If this fault is suspected, the clutch unit must be removed from the machine and the springs detached from the unit so that they may be inspected and, if necessary, renewed.

3 Place the clutch unit on a clean work surface ready for dismantling. Note, before attempting to dismantle the unit, that it contains a large spring which is held in compression between the sliding face of the driven pulley and the clutch backplate. It follows, therefore, that considerable care must be taken whilst removing the backplate retaining nut to prevent the force of the spring from stripping the holding threads of the retaining nut as it is being removed from the boss of the fixed face of the pulley and causing the backplate to be ejected upwards into the face of the person removing the nut.

4 Honda supply a set of special tools which may be used to hold the clutch backplate against the pulley unit whilst the retaining nut is removed and then allow the backplate to be released from the pulley unit under controlled conditions.

5 In the absence of the above tools it is possible to dismantle the clutch unit without them, provided care is taken and an assistant is available. The object is to partially compress the unit to take spring pressure off the retaining nut while it is unscrewed. This is necessary to avoid stripping the threads on the rather thin nut. The unit can be compressed sufficiently by placing it with the nut uppermost on the workbench and pressing down with the heel of both hands as shown in the accompanying photograph.

6 The nut should not be unduly tight and can be removed with a large adjustable spanner in most cases. If it seems particularly tight, it is preferable to use a large socket, ring or box spanner to avoid any risk of damage. If it is difficult to exert enough pressure on the unit to compress it, try rigging up an improvised compressor using either a large legged-puller or an engineers' vice and suitable spacers. If all else fails, take the unit to a Honda dealer who will be able to compress the spring and remove the nut. Once the nut has been released, the clutch boss can be lifted away, followed by the pulley spring and the moving face of the pulley.

7 Examine the surface for wear or damage. This is unlikely to occur unless the clutch shoes have been allowed to wear to the point that the backing metal has contacted the drum. Wear can be checked by measuring the inside diameter of the drum, which should not exceed 107.5 mm (4.23 in). The pulley spring free length should also be checked, and the spring renewed if it has compressed to the service limit of 88.8 mm (3.50 in) or less.

8 The three clutch shoes should be examined for wear, scoring or contamination of the friction surface by oil or grease. If contaminated or badly scored, the shoes must be renewed, as must the drum if this too is scored. Light contamination can be removed from the shoe linings using abrasive paper, but care should be taken to avoid inhaling the asbestos-based dust which will be produced. The friction material thickness of each shoe must not be below 2.0 mm (0.08 in) at any point. If it is at or below this limit, renew the shoes as a set, never singly.

9 To remove the shoes from the clutch boss, prise off the three circlips and washers which locate them, then gradually work the shoe assembly off the boss using screwdrivers. All three shoes must come off together, and should be eased evenly off the pivots. Once free of the hub the shoes and springs can be separated.

10 When fitting new shoes, mask the lining surface with masking tape or insulation tape to prevent contamination of the linings during fitting. Assemble the shoes and springs, then apply a smear of grease to each pivot. Check the condition of the three damper rubbers on the hub and renew them if damaged. Offer up the shoes, working them onto the pivots. This is not an easy task; the springs are quite strong and judicious levering will be required. Once in place, fit the washers and circlips to retain them.

11 Turning attention to the pulley assembly, slide off the headed seal collar to reveal the guide roller pins. These should be displaced and removed and the two halves of the pulley separated. Remove the oil seals and O-rings, taking care not to damage them. If these are worn or damaged in any way they must be renewed, but if great care is taken it should be possible to reuse them, provided that they are not distorted during removal.

12 Check the outer diameter of the fixed pulley face boss and the corresponding inner diameter of the moving pulley face boss for wear or damage. There should be no appreciable free play between the two, and this can be checked by direct measurement where suitable equipment is available. The relevant service limits will be found in the Specifications at the beginning of this Chapter. Check the guide roller pins and their grooves for wear or damage.

13 The two bearings in the centre of the fixed pulley half can be checked for wear or roughness without removal. If there are signs of excessive play or the bearing(s) feel rough or gritty when turned they should be removed. Drive out the inner roller bearing first, then remove the circlip and tap out the outer journal ball bearing. Using a socket as a drift, tap a new outer bearing into place with the sealed face outermost. Refit the circlip, and lubricate the gap between the bearings by applying 4 – 5 grams (0.14 – 0.16 oz) of grease. The inner bearing can now be installed with its numbered face outwards.

14 Grease the bore of the moving pulley half and fit the two O-rings in their grooves. Lubricate the lips of the oil seals and carefully press them into position, ensuring that they fit squarely. Assemble the two pulley halves and check that they move smoothly and evenly. Line up the guide pin slots and fit the two guide roller pins, retaining them by sliding the seal collar into place.

15 Reassemble the clutch and pulley unit, holding the spring compressed while the 28 mm retaining nut is secured. This should be tightened to 3.5 – 4.0 kgf m (25 – 29 lbf ft). Note that if using the official retaining nut wrench, a square is provided to allow a beam-type torque wrench to be used. This is, of course, some way from the nut centre, so to obtain the correct true torque figure, tighten until a reading of 3.3 – 3.8 kgf m (24 – 28 lbf ft) is obtained. Leave the protective tape in place on the clutch linings until the unit is installed, but remember to remove it before the drum is fitted.

24 Examination and renovation: transmission final reduction gearbox components

1 Carefully examine each of the gear pinions to ensure that there are no chipped or broken teeth. Gear pinions with this defect must be renewed as there is no satisfactory method of reclaiming them. Look also for signs of pitting on the load bearing faces of the teeth and for signs of severe wear which may have been caused through lack of lubrication. If damage or wear warrants renewal of any gear pinion, the complete gearbox assembly should be dismantled and laid out on a clean work surface. The same applies if any one of the gearbox shafts is seen to be worn or damaged.

2 If the decision is made to dismantle the gearbox assembly for the purposes of component renewal and further examination, each

component part should be either renewed, if seen to be obviously damaged, or washed in petrol and placed to one side ready for further examination and reassembly. Remember to observe the necessary fire precautions whilst using petrol to clean the gearbox components.

3　Lay both the new and cleaned components out on a sheet of clean paper or rag and proceed to carry out a further examination of the remaining original components as follows.

4　Fit the intermediate gear pinion to its shaft and feel for any excessive play between the two components. If any play found is thought to be unacceptable, then refer the problem to a Honda dealer who will be able to give an expert opinion and, if necessary, obtain the necessary replacement parts.

5　If it is necessary to remove the input shaft from the reduction gearbox cover, Honda recommend that it should be pressed out. Depending on how tight a fit the shaft is in its bearing, it may be possible to remove it by tapping it out. If this method is attempted, support the cover well above the surface of the workbench on wooden blocks. These must be placed as close as possible to the shaft to give the bearing boss support without risking damage to the cover casting. Use a copper or other soft metal drift to prevent damage to the shaft end, and avoid striking the shaft hard. If it does not tap out readily, abandon the attempt and have it removed by a press.

6　Check all bearings for signs of excessive free play or roughness when turned, indicating wear or damage of the bearing tracks or balls. If the input or output shaft left-hand (cover) bearings require renewal they can be removed using a 15 mm bearing extractor in conjunction with a slide-hammer. Alternatively, it is possible to tap the bearings out

using a suitable drift, having ensured that the bearing boss area is well supported on wooden blocks.

7　When dealing with the corresponding right-hand bearings, it will be noted that the input shaft bearing fits into a blind hole, and this means that a bearing extractor (12 mm in this case) and slide-hammer are essential. These tools can be obtained as Honda parts through official dealers, but it is also possible to hire slide-hammer and bearing extractor sets from most tool hire shops. Failing this, take the transmission case and cover to a Honda dealer for the work to be carried out. The output shaft (stub axle) bearing and seal can be removed using the slide-hammer arrangement or can be driven out. Whichever method is employed, remember that a tight bearing can be removed much more easily if the casing is warmed first in very hot water.

8　The new bearings should be tapped into place using a tubular driver and a selection of attachments which match the diameter of the outer race. Most owners can approximate this by using sockets of various sizes or by making tubular drifts from offcuts of steel tubing. Make sure that the bearings enter their bores squarely and seat fully. Fit new seals, having first lubricated the sealing lip with grease.

9　It will be noted that the idler shaft runs direct in the alloy of the transmission casing and cover, and it is conceivable that these surfaces will wear in time. If a new shaft fails to correct excessive free play, the casting concerned is theoretically unusable. In view of the cost of renewal of these castings it is suggested that a local engineering company is consulted with a view to having the bores reclaimed by boring them oversize and fitting bushes.

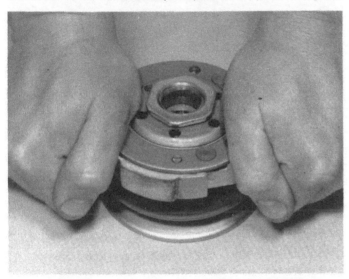

23.5 The clutch retaining nut can be unscrewed after compressing the spring as shown

23.6 Lift away the clutch assembly and the pulley spring

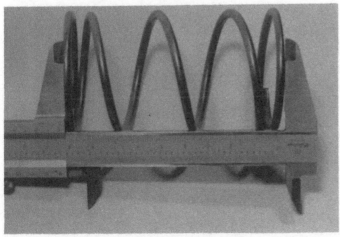

23.7 Measure pulley spring and renew if below limit

23.9a Clutch shoes are each retained by a circlip and washer

23.9b Once worked clear of the hub, separate shoes as shown

23.10 Check condition of the damper rubbers on hub prior to reassembly

23.11a Slide off the seal collar ...

23.11b ... to reveal the O-rings and guide roller pins. Displace and remove the pins ...

23.11c ... and separate the two pulley halves

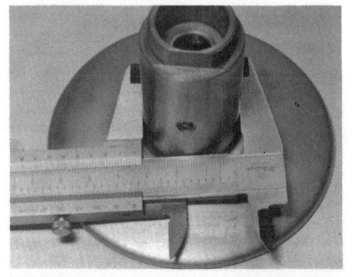

23.12 Check the diameter of the pulley boss using a vernier caliper

23.13a The inner bearing is of the needle roller type ...

23.13b ... whilst the outer bearing is a sealed ball race

24.7 Where a bearing is fitted in a blind bore (arrowed) a slide-hammer type bearing extractor will be needed

24.8a A large socket can be used to drive the new bearing home

24.8b Press the seal squarely into its bore and grease the seal lip

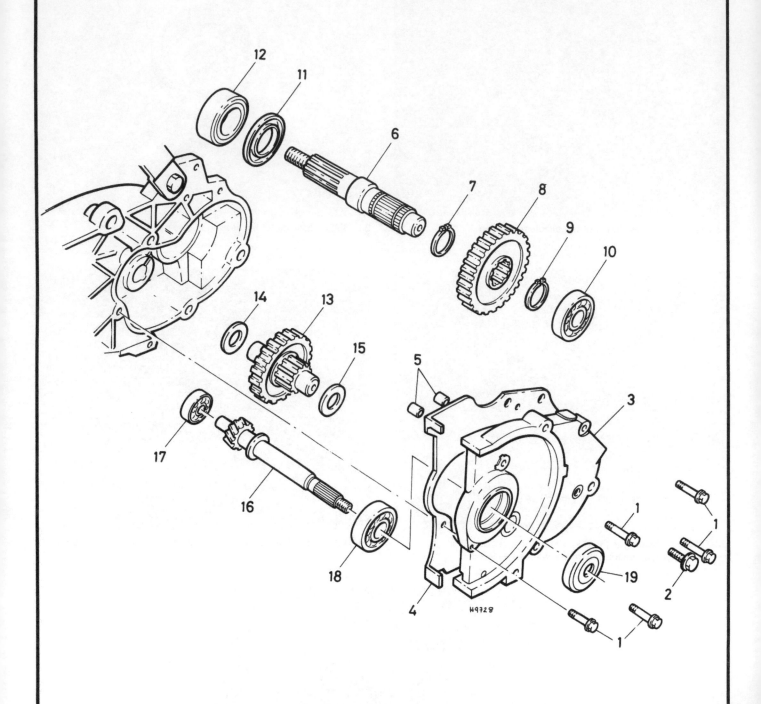

Fig. 1.9 Transmission final reduction gears

1	Bolt – 5 off	9	Circlip	15	Thrust washer
2	Level/filler plug	10	Output shaft right-hand bearing	16	Input shaft
3	Transmission cover			17	Input shaft right-hand bearing
4	Cover gasket	11	Oil seal		
5	Dowel pin – 2 off	12	Output shaft left-hand bearing	18	Input shaft left-hand bearing
6	Output shaft				
7	Circlip	13	Idler shaft	19	Oil seal
8	Final driven gear	14	Thrust washer		

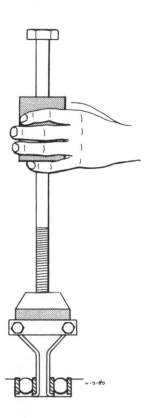

Fig. 1.10 Using a slide-hammer to remove a bearing from a blind hole

25 Examination and renovation: kickstart mechanism

1 Examine the teeth of the large kickstart pinion and the smaller driven pinion for wear or damage. Problems such as the mechanism slipping or jamming can often be attributed to worn or chipped teeth, and if such damage is discovered, renew the gears as a pair. The small driven gear also has three inclined teeth which engage with those at the centre of the front pulley when the kickstart lever is operated. Make sure that these teeth have not become rounded off or slipping may occur.

2 The friction spring which fits around the driven gear should hold the gear reasonably firmly. No specific figures are available to check the spring condition, but if it seems to turn on the gear too easily, either renew it as a precautionary measure or compare it with a new item.

3 The kickstart shaft should be a good fit in the transmission cover with no sign of side-to-side movement. It is unlikely that this will wear very quickly, especially if the electric starter is normally used. If badly worn, the normal course of action is to renew the cover and the shaft. This expense may be reduced by having the bore enlarged and a bush fitted, a job most engineering firms will be able to advise on and undertake.

26 Reassembling the engine/transmission unit: general

1 Before reassembly of the engine/transmission unit is commenced, the various component parts should be cleaned thoroughly and placed on a sheet of clean paper, close to the working area.

2 Make sure all traces of old gaskets have been removed and that the mating surfaces are clean and undamaged. One of the best ways to remove old gasket cement is to apply a rag soaked in methylated spirit. This acts as a solvent and will ensure that the cement is removed without resort to scraping, and the consequent risk of damage.

3 Gather together all of the necessary tools and have available an oil can filled with clean engine oil and one filled with clean gearbox oil.

Make sure that all the new gaskets and oil seals are to hand, also all replacement parts required. Nothing is more frustrating than having to stop in the middle of a reassembly sequence because a vital gasket or replacement has been overlooked.

27 Reassembling the engine/transmission unit: joining the crankcase halves

1 Before commencing work, make sure that the crankcase halves and the crankshaft assembly are clean and that a new crankcase gasket is to hand. Fit the crankshaft main bearings into their respective bores, driving them home squarely using a large socket as a drift against the bearing outer race. Where the right-hand main bearing is to be reused and has remained on the crankshaft, it can be fitted together with the crankshaft assembly. Unless the Honda crankcase assembly tooling is to be used (see below), fit new main bearing oil seals, having first greased their sealing lips. In the case of the right-hand seal, note that it must be positioned so that its outer face lies 9 mm (0.35 in) below the end of the boss in which it fits.

2 Honda recommend the use of two tools to fit the crankshaft and to close the crankcase joint. It is assumed that these are not available to the average owner, in which case proceed as described below. Place the transmission casing on the workbench with the crankcase side uppermost. Place wooden blocks beneath it to ensure adequate clearance for the crankshaft end when fitted. Make up a hardwood wedge which can be fitted between the two flywheels during crankshaft installation, and find a length of steel tubing having an internal diameter slightly greater than that of the crankshaft end.

3 Offer up the crankshaft, pushing it home as far as possible by hand. Turn the crankshaft so that the crankpin is diametrically opposite the crankcase mouth. Insert the wedge so that it supports the flywheels opposite the crankpin. This will counteract any tendency for the assembly to become distorted during fitting. Place the tubular drift over the crankshaft end and tap the assembly home. Do not strike the crankshaft hard or damage may result. If it seems excessively tight, check that it is entering the main bearing squarely.

4 Fit the locating dowels into their bores and place a new gasket in position. Lower the crankcase right-hand half over the crankshaft, pushing it as far as it will go by hand. Check that the wedge is still in position, then using a tubular drift, tap the crankcase joint closed. Once the mating surfaces have closed, fit the retaining bolts and tighten them provisionally in a diagonal sequence. If the crankshaft is still under some tension and turns stiffly this can be corrected by tapping each end of the crankshaft in turn, using a soft-faced mallet, until the crankshaft settles and turns freely. Once this has been checked, remove the wooden wedge from between the flywheels and tighten the retaining bolts in a diagonal sequence to 0.8 – 1.2 kgf m (6 – 9 lbf ft).

27.4 Fit the crankshaft into the left-hand casing half, then fit a new gasket

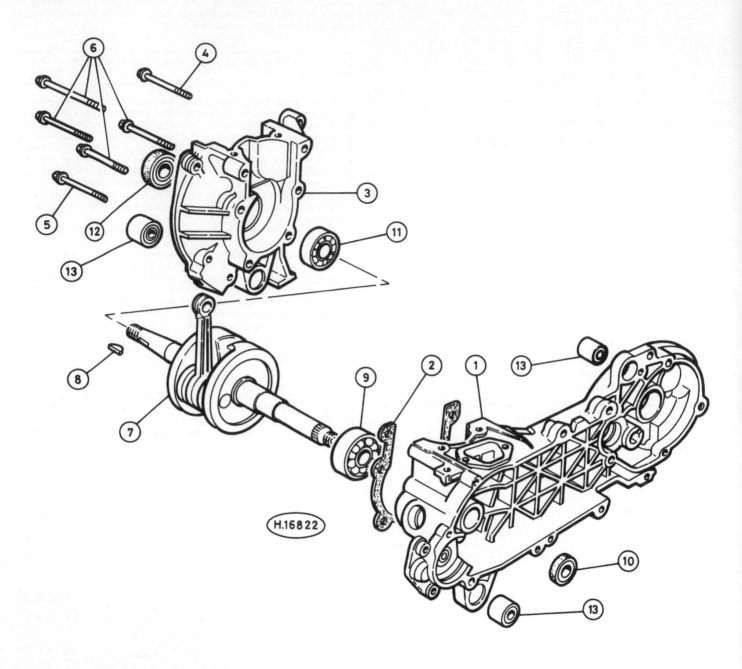

Fig. 1.11 Crankcase

1	Left-hand crankcase/	5	Bolt	10	Left-hand oil seal	
	transmission housing	6	Bolt – 4 off	11	Right-hand main bearing	
2	Gasket	7	Crankshaft	12	Right-hand oil seal	
3	Right-hand crankcase	8	Woodruff key	13	Mounting bush – 3 off	
4	Bolt	9	Left-hand main bearing			

28 Reassembling the engine/transmission unit: refitting the final reduction gearbox components

1   Check that the reduction gearbox bearings and oil seals are fitted correctly, then install the idler gear assembly with its thrust washer. Lubricate the output shaft (stub axle) and fit it through its bearing and seal. Place the second thrust washer over the end of the idler shaft. If

removed, fit the output shaft gear circlip.
2   Check that the two locating dowels are in position, then fit a new gasket. Pay particular attention to the two foam sealing pads at the forward ends of the gasket. These must bend around the mating face to be trapped by the cover (see photograph). Offer up the cover, complete with the input shaft, and check that the gearbox turns normally. Fit the five retaining bolts, tightening them evenly and progressively to preclude any risk of distortion.

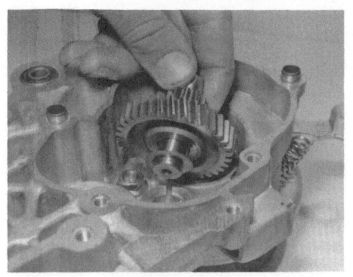

28.1a Fit the idler shaft, noting the thrust washer

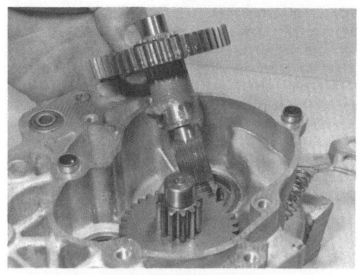

28.1b Fit the output shaft (stub axle) through its bearing ...

28.1c ... and fit the second thrust washer to the idler shaft

28.2a Fit a new gasket to the transmission casing and lower the cover into position

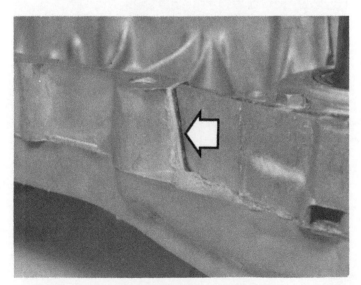

28.2b Make sure that the sealing pads locate as shown

29 Reassembling the engine/transmission unit: refitting the transmission components

1   Place the starter pinion assembly in its recess at the front of the transmission casing. Fit the inner part of the front pulley assembly over the crankshaft end, leaving the outer pulley half off at this stage.

2   Loop the drive belt around the rear pulley assembly. Squeeze the pulley halves apart against spring pressure and pull the loop of the drive belt in against the pulley centre to hold it open (see photograph). Place the rear pulley over the end of the input shaft, still holding the belt to hold the pulley halves apart. Loop the front of the belt over the front pulley and place the outer pulley half in position. Fit the plain washer, then fit the retaining nut finger tight before releasing the belt and the rear pulley halves.

3   Using the method employed during removal, lock the crankshaft and secure the front pulley nut. If masking tape was applied to the clutch shoes, peel this off, then place the clutch drum over the input shaft. Fit the retaining nut and tighten it to 3.5 – 4.0 kgf m (25 – 29 lbf ft), using the holding tool to prevent it from turning.

29.1a Place the starter pinion assembly into its recess in the transmission casing

29.1b Fit the inner half of the front pulley over the crankshaft end

29.2a Loop the belt around the rear pulley to hold the pulley halves apart, then fit it over the input shaft

29.2b Fit the outer half of the front pulley, then release the belt

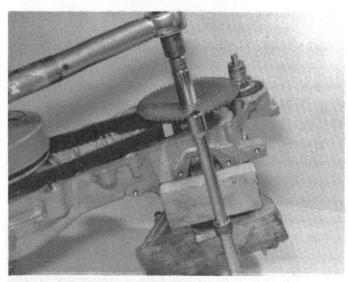

29.2c Lock the crankshaft and tighten the front pulley nut to the prescribed torque setting

29.3 Remove masking tape from shoes, if fitted, then fit the clutch drum and retaining nut

30 Reassembling the engine/transmission unit: refitting the kickstart mechanism and transmission cover

1   Lay the kickstart return spring over the raised boss inside the cover, hooking the end over the anchor pin. Grease and fit the kickstart spindle sleeve, then slide the kickstart shaft into position. As the shaft is slid into place, hook the free end of the return spring over the notch in the edge of the gear.

2   Lower the kickstart driven gear into place, ensuring that the friction spring is hooked over its locating pin. It will be necessary to turn both gears so that they mesh correctly; when installed, the large kickstart pinion should be stopped by a plain section in its teeth jamming on the driven gear. The accompanying photographs show the main gear with the spring pressure released, and then tensioned and in mesh with the driven gear. Finally, turn the cover over and fit the plain washer and circlip.

3   Before the cover is refitted, it is worth checking that the kickstart mechanism operates normally. Temporarily refit the cover, securing it with three or four bolts, and refit the kickstart lever. Operate the kickstart lever by hand a few times to ensure that it engages normally.

4   Check that the mating surfaces of the transmission housing and cover are clean, then place a new gasket in position. Offer up the cover and fit the retaining bolts. Note that the third bolt from the back on the lower edge of the casing also retains the guide clip for the rear brake cable.

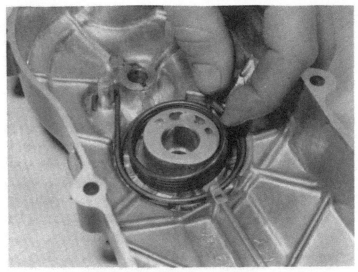

30.1a Place kickstart spring in position as shown ...

30.1b ... and slide the greased spindle sleeve into bore in cover

30.1c Fit the kickstart shaft, hooking the free end of the spring over the notch in the gear

30.1d Fit the plain washer over the shaft end ...

30.1e ... and secure it with its circlip

30.2a Turn kickstart gear and install the driven gear, noting position of friction clip

30.2b Gears should look like this when meshed correctly

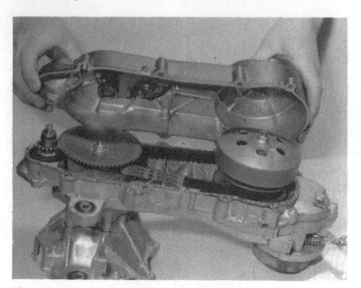

30.4a Fit the transmission cover, using a new gasket ...

30.4b ... and tighten retaining bolts evenly

31 Reassembling the engine/transmission unit: refitting the piston, cylinder barrel and cylinder head

1   Position the crankcase unit upright on the work surface and move the connecting rod to its top dead centre (TDC) position. Pack a piece of clean rag between the sides of the connecting rod and the crankcase mouth to prevent any component parts from falling into the crankcase.

2   Fit the small-end bearing into the eye of the connecting rod and lubricate it thoroughly with clean engine oil. Position the piston over the connecting rod so that the 'EX' mark on the piston crown is facing the front of the engine. Push the gudgeon pin into position. The pin should be a firm sliding fit but if it proves to be tight then warm the piston in hot water to expand the metal around the gudgeon pin bosses.

3   Always use new circlips to retain the gudgeon pin in position and double check that each clip is correctly located in the piston boss groove. Note that each circlip must be fitted so that its gap is well away from the cutouts in the sides of the gudgeon pin hole. A circlip that is allowed to work loose will cause serious damage to both the cylinder bore and piston.

4   Check that the piston rings have not been disturbed from their set positions and then remove the rag from the crankcase mouth. Place a new cylinder barrel base gasket in position on the crankcase. Lubricate the big-end, the piston rings and the cylinder bore with clean engine oil. Grip the piston in one hand whilst pressing the rings into their grooves. The cylinder barrel can now be lifted into position over the crown of the piston and guided carefully down over the piston rings. There is a generous lead-in on the base of the cylinder bore which serves to make this operation reasonably easy. With the piston pushed fully into the bore, align the cylinder barrel with the crankcase mating surface and push it into position.

5   Wipe any excess engine oil off the upper surface of the cylinder barrel and place the new cylinder head gasket in position on the barrel. Fit the cylinder head and push the four retaining bolts into position through the cylinder head and barrel. Do not force these bolts. If difficulty is experienced in fitting them, recheck that the cylinder barrel, cylinder head and their respective gaskets are all in correct alignment with the surface of the crankcase mouth. Tighten the bolts, finger-tight at first and then evenly and in a diagonal sequence until the recommended torque loading of 7.0 – 9.0 lbf ft (0.9 – 1.2 kgf m) is reached. Finally, temporarily plug the spark plug hole with a wad of clean rag to prevent the ingress of any contamination into the combustion chamber. The spark plug should not be fitted until the engine unit is refitted into the frame.

31.2a Lubricate and fit the small-end bearing

31.2b Piston must be fitted with 'EX' mark facing the exhaust port

31.2c Slide the gudgeon pin into place to retain piston

31.3 Retain the gudgeon pin using **new** circlips

31.4a Check that ring ends engage over the pegs in ring grooves

31.4b Remove rag from crankcase and fit a new base gasket

31.4c Guide rings into bore, pushing the barrel straight down without twisting it

31.5a Fit a new cylinder head gasket ...

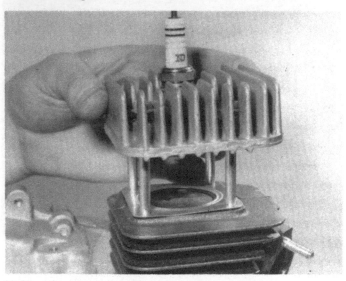

31.5b ... then install the cylinder head and retaining bolts

31.5c Tighten the bolts evenly and progressively to the prescribed torque setting

32 Reassembling the engine/transmission unit: refitting the starter motor and oil pump

1   Check that the O-ring at the end of the starter motor casing is in position and in good condition, then apply a thin film of grease to ease installation. Offer up the motor, pushing it fully home in its recess, and fit the two retaining bolts to secure it. Tighten the bolts evenly.

2   The oil pump is fitted in a similar fashion, again making sure that the O-ring is in place and lubricated. Offer up the pump, turning it as necessary to ensure that the driven gear engages correctly. Once in position, the pump is secured by a retainer plate held by one of the cylinder cowling screws. If this is not in place yet, leave this operation until the cowlings are fitted (see Section 34).

33 Reassembling the engine/transmission unit: refitting the reed valve and intake adaptor

1   Check that the reed valve body and the crankcase and intake adaptor mating faces are clean and that all traces of old gasket have

been removed. Remove any rag used to plug the inlet port when the reed valve was removed. It is important to avoid any risk of air leaks around the reed valve assembly or erratic running will result. Place a new gasket on either side of the reed valve, holding them in place with a smear of grease.

2   Offer up the reed valve unit, noting that the valves project down into the inlet port. The intake adaptor can be fitted next, its two retaining bolts passing through the reed valve case and into the crankcase. Tighten the bolts evenly to 0.8 – 1.2 kgf m (6 – 9 lbf ft) only, to prevent distortion of the mounting flange. Reconnect the oil delivery pipe between the pump and the intake adaptor.

34 Reassembling the engine/transmission unit: refitting the flywheel generator and engine cowlings

1   If not already in place fit the cylinder cowling, noting that the front mounting bolt also holds the oil pump retainer. Make sure that the retainer locates the oil pump correctly, then tighten the cowling bolts.

2   Offer up the generator stator and ignition pickup assembly, ensuring that the wiring passes through the casing cutout. The stator and pickup are each retained by two mounting screws.

3   Fit the Woodruff key into the slot in the crankshaft end and offer up the generator rotor. Fit the rotor retaining nut finger-tight, then lock the crankshaft and tighten the nut to 3.5 – 4.0 kgf m (25 – 29 lbf ft).

4   Place the plastic cooling fan over the rotor and secure it with its two retaining screws. The fan cowling can now be fitted. If it was removed during the overhaul, refit the small rubber flap on the cylinder cowling, arranging it so that it will deflect water and mud from the rear wheel clear of the carburettor area.

35 Reassembling the engine/transmission unit: refitting the carburettor

1   Check that the O-ring seal is in place in its groove on the carburettor flange and that it is in good condition. Offer up the carburettor, together with the heat insulator, and fit the two mounting bolts finger-tight. Note that the bolts are of differing length; the longer bolt is fitted on the inner edge of the intake adaptor.

2   Tighten the retaining bolts evenly to avoid distorting the mounting

flange to the recommended torque setting of 0.9 – 1.2 kgf m (7 – 9 lbf ft). Note that the throttle valve, carburettor top, fuel and vacuum pipes, and the choke leads must be connected after the engine/transmission unit has been installed.

36 Reassembling the engine/transmission unit: refitting the rear wheel and exhaust system

1   If the rear wheel was removed in the course of the overhaul it can be refitted at this stage. Check that the rear brake linings are clean and free from grease marks and remove masking tape if this was applied to protect them. Offer up the wheel, ensuring that it locates over the stub axle splines. Fit the retaining nut and tighten it to 8.0 – 10.0 kgf m (58 – 72 lbf ft).

2   Fit a new exhaust port seal, then offer up the system. Fit the two exhaust port nuts finger-tight, followed by the two silencer mounting bolts. Finally, tighten both nuts and bolts evenly and firmly.

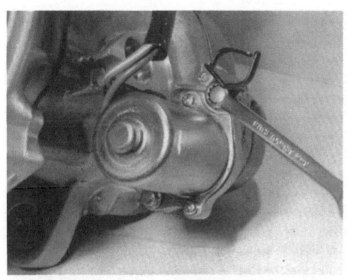

32.1 Starter motor is retained by two bolts

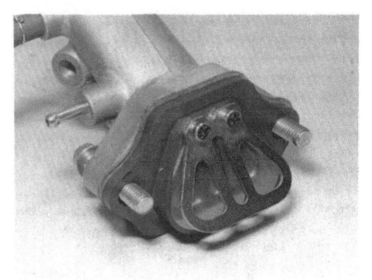

33.2 Install the reed valve and the inlet adaptor

34.2 Refit the generator stator and ignition pickup, then fit the Woodruff key in crankshaft slot

34.3a Fit the rotor over the crankshaft end ...

34.3b Hold the rotor, using home-made tool, and tighten rotor nut

35.1 Install the carburettor body on the inlet adaptor

36.1 Fit rear wheel and secure with new locknut

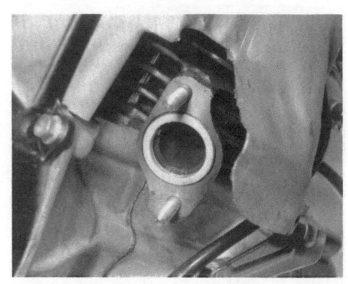

36.2a Place a new sealing ring in the exhaust port

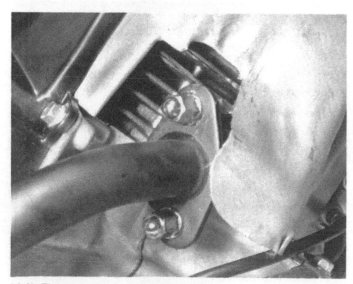

36.2b Tighten the exhaust flange nuts first ...

36.2c ... followed by the silencer mounting bolts

37 Reassembling the engine/transmission unit: checking the engine mounting

1   Before the engine/transmission unit is refitted in the frame, the main engine mounting should be checked. This is a pivoting bracket supported on a rubber bush, inserted into the frame tube. If all is well there should be a little up and down movement controlled by the rubber bush, but no free play in the pivot.

2   If play is evident, remove the pivot bolt and withdraw the mounting to reveal the bush. This can be examined, and if necessary renewed, before the engine mounting is refitted and the engine/transmission unit installed.

38 Refitting the engine/transmission unit into the frame

1   It is worth checking at this stage that nothing has been omitted during the reassembly sequences. It is better to discover any left-over components at this stage rather than just before the engine is due to be started.

2   Refitting of the engine/transmission unit into the frame is, generally speaking, a direct reversal of the removal sequence. As with removal, it may be found advantageous to have an assistant present to steady the machine whilst the engine unit is eased into position.

3   Commence refitting the engine unit by selecting the unit pivot bolt and applying a light coating of a high melting-point, lithium based grease to its shank. Doing this will ensure that the bolt does not become corroded to the centre of the bonded rubber bushes in the unit mounting plate, thus making extraction of the bolt at a later stage relatively trouble free.

4   Carefully ease the engine unit into position in the frame so that the unit mounting plate is aligned correctly between the two frame plates. Working from the left-hand side of the machine, push the pivot bolt through the mounting plates and tap it fully home with a soft-faced hammer. Fit the nut and washer to the bolt and tighten the nut to a torque loading of 25 – 33 lbf ft (3.5 – 4.5 kgf m).

5   Reconnect the rear suspension unit at its lower mounting point, tightening the mounting bolt to 2.0 – 3.0 kgf m (14 – 22 lbf ft). Check that the engine mounting bracket bolt has been tightened securely, if it was disturbed.

6   Connect the vacuum pipe between the lower stub on the fuel valve and the stub on the inlet adaptor. Connect the fuel pipe between the side stub on the fuel valve and the carburettor. Check that both pipes are secured with their retaining clips. Remove the temporary plug from the end of the oil tank feed pipe and allow any trapped air to be displaced before pushing it onto the oil pump stub. Note that the lubrication system must be checked and bled before the machine is used.

7   Coat the throttle valve with light oil and introduce it into the carburettor body, making sure that the throttle valve is fitted the right way round (it locates over a guide pin). As the valve is inserted, check that the needle enters the needle jet squarely.

8   Screw down the carburettor top and check that the throttle cable free play is set correctly. This is measured at the edge of the twistgrip flange and should be 2 – 6 mm ($^1/_8$ – $^1/_4$ in). If adjustment is required, slide back the protective sleeve on the adjuster. This is located at the upper end of the cable, where it enters the right-angled bend. Slacken the locknut and set the cable adjustment, then secure the locknut and slide the sleeve back into position.

9   Reconnect the rear brake cable at the wheel end, ensuring that the cable passes through the guide clip under the transmission casing. Set the cable free play to give 10 – 20 mm ($^3/_8$ – $^3/_4$ in) movement at the lever end. Check that the brake and the locking lever work correctly before moving on.

10 Refit the air filter casing, ensuring that it locates over the carburettor intake. Secure the casing using the two remaining transmission cover screws. Open the cover of the ignition coil housing, on the top of the air filter and slide the coil into place. Check that the coil wiring is routed correctly, then shut the cover.

11 If removed, refit the spark plug after removing the wad of rag used to plug its hole, screwing it home finger-tight, then tightening it by $^1/_4$ to $^1/_2$ a turn using a plug spanner. Reconnect the plug cap and make sure that the dust seal around it fits correctly against the cylinder cowling. Note that a clip on the air filter casing holds the end of the rubber flap which extends from the cowling to protect the carburettor. Hook the flap into place, then clip the HT lead in behind it.

12 Reconnect the wiring at the various connectors located on the left-hand side of the machine, paying attention to the wiring colour-coding. Slide the protective sleeve over the connectors and clip it into place on the frame. Reconnect the battery and check that the electrical system functions normally, bearing in mind that the headlamp circuit does not operate unless the engine is running.

13 Remove the filler/level plug from the final reduction gearbox and top it up to the level of the hole with SAE 10W/40 motor oil. The casing holds about 90 cc (3.2 Imp fl oz) of oil, and a means of getting it into the casing will need to be devised in view of the filler orifice position. A syringe is ideal, but failing that an old, but clean, fork oil or gear oil container could be used. Make sure that the oil is just level with the filler plug threads, then refit and tighten the plug.

38.4 Manoeuvre engine unit into frame and refit pivot bolt

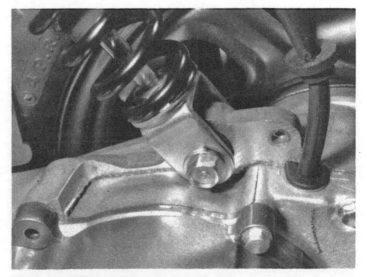

38.5 Fit suspension unit lower mounting bolt

38.6a Reconnect the fuel pipe (A) and vacuum pipe (B)

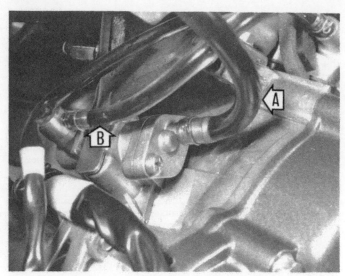

38.6b Fill oil pipes to exclude air, then connect the oil tank pipe (A) and delivery pipe (B)

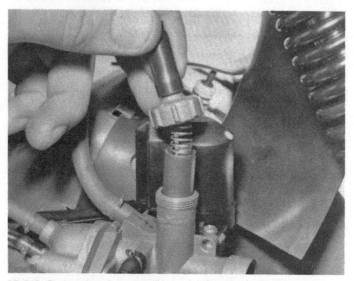

38.7 Refit throttle valve assembly and tighten carburettor top

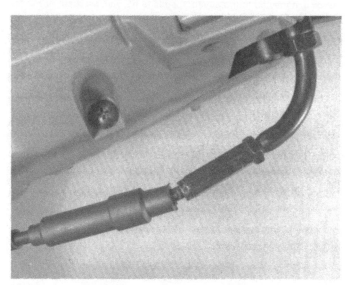

38.8 Adjust throttle cable play at adjuster below twistgrip

38.9 Fit rear brake cable into guide clip under transmission casing

38.10a Refit air filter casing, making sure that carburettor engages as shown

38.10b Place ignition coil into housing and close the lid

38.11 Hook rubber flap onto air filter case, then clip HT lead as shown

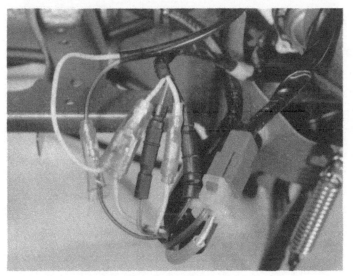

38.12 Reconnect wiring and clip to the frame

38.13 Top up transmission final reduction gearbox until oil is level with filler plug threads

39 Starting and running the rebuilt engine

1   Make a final check around the machine to ensure that the wiring, all control cables and pipes have been connected properly. Apply the rear brake and lock it on. Switch on the ignition and operate the starter button. It is quite likely that the engine will be reluctant to start at first, particularly if the carburettor float chamber is empty; it will take a while for it to fill, and this can only occur while the engine is turning. Care should be taken not to discharge the battery. If necessary, crank the engine for a few minutes, then allow the battery to recover for a minute or two before repeating the attempt.

2   If the engine shows no sign of starting, check that the fuel pipe and vacuum pipe are connected correctly, and that there is sufficient fuel in the tank. In persistent cases it may be worth removing the spark plug to check that oil in the cylinder has not fouled the plug electrodes. When the engine finally starts, it will almost certainly emit a fair amount of smoke from the exhaust. This is quite normal and consists of excess oil applied during assembly. The smoking should subside after a few minutes.

3   The oil used during assembly will ensure adequate initial lubrication, but the engine lubrication system must be checked soon after the initial start is made. Observe the pipe from the oil tank to the pump and the pipe from the pump to the intake adaptor. Both should be filled with oil and free from air bubbles. Any small bubbles should work through without too much of a problem, but large amounts of air can prevent the system from operating.

4   If there is a significant amount of air in either pipe, stop the engine and disconnect the pipe, then force oil through the pipe to displace the air using a syringe or a pump-type oil can. Leave the intake adaptor end of the delivery pipe disconnected and place some rag beneath it, then start the engine. Oil should start to emerge from the pipe within a minute, but if it fails to do so, prime the pipes again and repeat this check. Once all is well, reconnect the delivery pipe and wipe up any spilled oil.

5   When you are satisfied that the lubrication system is operating correctly and that the machine is running normally, stop the engine and refit the body panels and rack (see Routine Maintenance). The machine can be taken for a test ride, having first gone through the pre-ride check list shown at the beginning of Routine Maintenance, and paying particular attention to areas like the brakes, which may have been disturbed during the overhaul.

6   A rebuilt engine must be allowed to settle in gradually especially where new engine components, such as a piston and rings, have been fitted. During the first few hundred miles, the various moving parts will have to bed in, and it is essential not to work the engine too hard during this initial period. Careful running-in will allow a good seal to be formed between the piston rings and bore, and will prevent the risk of engine seizure during this process. The engine can be allowed to run reasonably briskly, but beware of over-revving or allowing the engine to labour up steep hills.

7   After the initial test ride, check the machine over carefully, for signs of oil or fuel leaks or loose components or fasteners. Rectify any such problems immediately. Never add a little 'extra' oil to the fuel in the mistaken belief that this will improve engine lubrication; all this will do is effectively weaken the fuel/air mixture and make the engine more likely to seize. Likewise, beware of any other condition which may affect the mixture, such as damage or leaks in either the exhaust or air filter systems.

Chapter 2 Fuel system and lubrication

Refer to Chapter 7 for information relating to the SA50 Met-in

Contents

General description 1	Carburettor: settings 10
Fuel tank: removal and refitting 2	Reed valve: examination and renovation 11
Fuel tank: flushing 3	Air filter element: removal and cleaning 12
Fuel tap: fault diagnosis, removal and refitting 4	Exhaust system: removal and refitting 13
Fuel and vacuum pipes: examination and renovation 5	Exhaust system: decarbonising and refinishing 14
Cold start device: fault diagnosis, removal and refitting ... 6	Oil tank: removal and refitting 15
Carburettor: removal and refitting 7	Oil tank filter: removal and cleaning 16
Carburettor: examination and renovation 8	Oil pump: removal and refitting 17
Carburettor: adjustment 9	Oil pump and pipes: bleeding 18

Specifications

Fuel grade.. Unleaded or low-lead, minimum octane 91 (RON)

Fuel tank

Capacity .... 4.0 litres (0.88 Imp gallon)

Carburettor

Type	Slide, automatic cold start
Venturi diameter	14 mm (0.55 in)
Identification No:	
NE50M-F...	PA04H-B
NE50TH..	PA04Q-B
NB50M-F...	PA04G-A
Main jet:	
NE and NB50M-F..	80
NE50TH...	82
Float level	12.2 mm (0.48 in)
Pilot air screw setting (turns out)	$2^{1}/_{8}$
Idle speed	1800 ± 100 rpm
Throttle free play	2 – 6 mm ($^{1}/_{8}$ – $^{1}/_{4}$ in)

Engine lubrication

Type	Pump fed from oil tank
Oil grade	Honda 2-stroke injector oil or equivalent
Tank capacity	0.8 litre (1.4 Imp pints)

1 General description

The fuel system is unusually sophisticated for a small capacity two-stroke machine, featuring a number of rather complex devices which serve to remove some manual controls. The first of these devices is an automatically operated choke control valve which obviates the need for the normal hand operated, cable/lever mechanism. The second device is a vacuum operated fuel tap which is brought into operation directly the engine is turned, by operating the starter motor or the kickstart. This tap replaces the usual hand operated type and therefore obviates the need to remember to turn the tap on or off. The carburettor is of the normal slide type but features the adaptors and

drillings necessary for operation of the choke control valve.

The fuel/air mixture is admitted to the engine via a reed valve assembly which allows more accurate timing of the mixture than conventional piston porting alone. The valve is entirely automatic in operation, opening and closing according to changes in pressure inside the crankcase.

The engine is lubricated by a pump-fed total-loss oil system. The pump is mounted at the front of the crankcase, drawing oil from a plastic tank below the seat. Oil from the pump is injected into the inlet adaptor, from where it is drawn into the engine with the fuel/air mixture. Pump operation is automatic, delivery varying according to engine speed. There is no connection to the throttle twistgrip as on most similar systems.

2 Fuel tank: removal and refitting

1 The fuel tank is located below the seat. Start by removing the rack and the two side panels as described in Routine Maintenance, then free the seat by removing the two nuts which secure it to the front of the tank.

2 When working on the fuel system note the following safety precautions. Take great care to place the machine in a well ventilated space where there is no chance of a build-up of fuel vapour occurring with the subsequent risk of a fire or explosion. Drain any fuel from disconnected pipes into a clean metal container that is equipped with a properly sealed lid; never use an open or plastic container. Store both this container and the removed tank in a well ventilated open space. Never smoke or expose the machine to any kind of naked flame whilst removing or fitting the fuel tank.

3 Detach the fuel and vacuum pipes from the fuel tap by releasing their spring clips and pulling them off their retaining stubs. Note that the pipe attached to the horizontal stub of the tap is the one that supplies fuel to the carburettor. Drain any fuel retained in this pipe into a container. If fuel continues to issue from the fuel tap, then the tap is defective and must be serviced as described in Section 4 of this Chapter.

4 Trace the electrical wires from the float switch fitted to the top of the tank to their nearest push connectors. Unplug these connectors and place them on the tank top. With the tank properly supported, remove each of its four mounting bolts and then lift the tank clear of the frame.

5 Fitting the tank is a direct reversal of the removal procedure. Inspect the fuel and vacuum pipes for deterioration or damage before reconnecting them and renew each one as necessary. Note that the electrical wires to the float switch are colour coded to avoid confusion whilst reconnecting. Once the tank is fitted in the machine, start the engine and allow it to run at tick-over speed whilst carrying out a thorough check around the pipe connection points for any signs of fuel leakage. If a leak is found, it must be cured before the machine is ridden, otherwise there is a considerable risk of fire resulting in serious injury to the rider.

3 Fuel tank: flushing

1 The fuel tank may need flushing out occasionally to remove any accumulated debris which inevitably builds up over the years. This is especially true if water has contaminated the fuel, as this can cause persistent and annoying running problems as it gets drawn into the carburettor.

2 Flushing is best done by first removing the tank, as detailed in the preceding Section, and then draining the tank of any contaminated fuel. Any debris or water may now be cleaned out by flushing the tank with clean petrol. Note that this operation must be undertaken outdoors and away from naked flames or lights, otherwise serious personal injury may result from the the ignition of the fuel vapour.

4 Fuel tap: fault diagnosis, removal and refitting

1 To gain access to the fuel tap for examination or removal it is first necessary to detach the left-hand side panel as described in Routine Maintenance. Before disconnecting any part of the tap assembly, take note of the following safety precautions. Take great care to place the machine in a well ventilated space where there is no chance of a build-up of fuel vapour occurring with the subsequent risk of a fire or explosion. Drain any fuel into a clean metal container that is equipped with a properly sealed lid; never use an open or plastic container. Store this container in a well ventilated open space. Never smoke or expose the machine to any kind of naked flame whilst draining fuel.

2 If difficulty is experienced in starting the engine and fuel starvation is suspected, detach the vacuum pipe from the intake adaptor and the fuel feed pipe from the carburettor, having first removed the carburettor shield to gain access. Have a container nearby in which to catch any fuel that may drain from the fuel pipe.

3 Test the operation of the fuel tap by placing the end of the fuel pipe in the container and then applying suction to the end of the vacuum pipe to cause the diaphragm in the tap to move downwards against its return spring. Once this diaphragm is moved, fuel will flow from the tank, through the tap and out of the end of the fuel pipe. If this does not happen, the fuel starvation can be attributed to one the following causes:

    A stuck fuel tap diaphragm
    A blocked vacuum pipe
    A blocked fuel pipe
    A blocked fuel tap filter

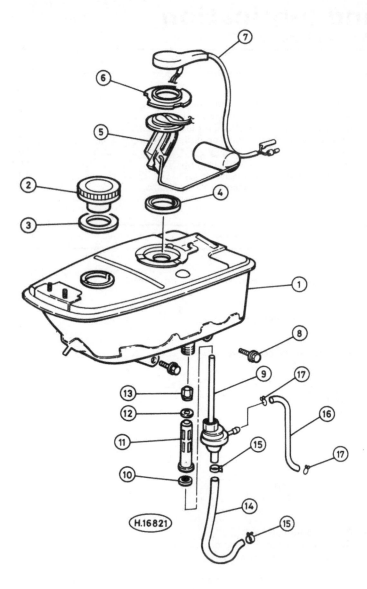

H.16821

Fig. 2.1 Fuel tank

1	Fuel tank	10	Seal
2	Filler cap	11	Fuel filter
3	Gasket	12	Clip
4	Seal	13	Air filter
5	Fuel level sender	14	Vacuum pipe
6	Retaining plate	15	Pipe clip – 2 off
7	Sender wiring	16	Fuel pipe
8	Bolt – 4 off	17	Pipe clip – 2 off
9	Fuel tap		

Only a gentle amount of suction need be applied to the end of the vacuum pipe to cause operation of the tap diaphragm. This suction may be applied by placing the pipe in one's mouth and sucking. Some care should be taken whilst doing this to ensure that there is no fuel present in the vacuum pipe as would be the case if the tap diaphragm were split. An alternative method of applying suction is to use a foot or hand pump.

4 If operation of the tap is found to be at fault, then detach both the fuel and vacuum pipes from the tap and check each one for any sign of a blockage in its bore. Clear any blockage found by blowing through the pipe with compressed air. Reconnect the pipes to the fuel tap and secure them with their spring clips.

5 It is possible to free a tap diaphragm that is stuck by applying a jet of compressed air to the end of the fuel pipe. This will cause the diaphragm to be pushed down against its spring, thereby freeing it from the closed position. Take care not to use too great a pressure of air when doing this, otherwise the diaphragm will be permanently damaged. If the diaphragm will not free or if fuel is seen to issue from the end of the vacuum pipe, the tap is defective and should be renewed.

6 To gain access to the fuel tap filter for the purposes of inspection and cleaning, it is necessary to completely drain the tank of fuel and then remove the fuel tap from the tank. Draining the tank can be achieved by either applying suction to the vacuum pipe to allow fuel to pass from the tank into a container or by removing the tank and inverting it to allow fuel to be drained out.

7 With the tank thus drained, unscrew the locknut which serves to retain the fuel tap in position and carefully pull the tap out of its location whilst taking care not to cause damage to the filter stack. The filter can now be removed from the tap together with its base gasket.

8 Clean the fuel filter by rinsing it in clean fuel. Any stubborn traces of contamination can be removed by gently brushing the filter with a soft-bristled brush which has been soaked in clean petrol; a used toothbrush is ideal. Complete the cleaning process by blowing the filter clear with a jet of compressed air. Remember to take the necessary fire precautions whilst carrying out this cleaning procedure and always wear eye protection against any fuel that may spray back from the brush or air jet. On completion of cleaning, closely inspect the filter for any splits or holes that will allow the passage of sediment through it and into the carburettor. Renew the filter if it is in any way defective.

9 Inspect and, if necessary, renew the small base gasket of the filter. Carefully clean the small air filter with a jet of compressed air and carry out a general inspection of the fuel tap for any signs of damage. Make sure that the air pipe is clear of contamination and fit the fuel filter with its base gasket over it. Push the fuel filter down into the centre of the tap locknut to seat it and then fit to the air pipe the support clip for the air filter. This clip must be located so that it is 17 mm (0.67 in) from the top of the air pipe. Fit the air filter and then carefully relocate the fuel tap in the tank. Hold the tap so that its fuel and vacuum pipe retaining stubs are correctly positioned and then tighten the locknut. Pour a small amount of fuel into the tank and check for any leakage of fuel from around the tap to tank joint. If fuel is seen to leak from this joint and the leak cannot be stopped by further tightening of the nut, it will be necessary to remove the tap and relocate or renew the fuel filter base gasket. Do not overtighten the tap locknut.

10 Finally, after having fitted the fuel tap and reconnected the fuel and vacuum pipes, refill the fuel tank and start the engine. Allow the engine to run at tick-over speed whilst carrying out a thorough check for leaks at all of the disturbed connection points. If a leak is found, it must be cured before the machine is ridden, otherwise there is a considerable risk of fire, resulting in serious injury to the rider.

5 Fuel and vacuum pipes: examination and renovation

1 The condition of the fuel and vacuum pipes connected to the fuel tap should be checked periodically. Synthetic rubber pipes are resistant to deterioration, but may eventually develop leaks where they push over their respective stubs. This can be corrected by disconnecting the pipe and slicing off the worn end. For obvious reasons this cannot be repeated many times before renewal becomes necessary.

2 Always obtain replacement pipes of the correct type, especially where identification marks are present. Replacement pipes of different materials should be avoided, and natural rubber tubing must be avoided at all costs. This latter material is attacked by fuel and will disintegrate internally, blocking the carburettor jets with a thick rubbery sludge.

6 Cold start device: fault diagnosis, removal and refitting

1 The Vision models are equipped with a fully automatic cold start device, known by Honda as an auto bystarter. The unit consists of a conventional plunger-type valve operated by a small solenoid. It is housed in a black plastic casing mounted on top of the carburettor. When the ignition is first switched on, the plunger is raised, allowing a fuel-rich mixture to be fed to the engine to permit easy starting. After about five minutes of running, the valve is shut off, leaving the engine to run on its normal mixture.

2 The system is entirely unobtrusive in operation, and any fault will soon become apparent to the rider. If the valve fails to open when the engine is cold, starting will be difficult or even impossible, with a marked tendency towards stalling until the engine reaches full operating temperature. Conversely, if the valve becomes stuck open, the engine will start normally, but will run irregularly when hot, with a noticeable increase in exhaust smoke.

3 If a fault is suspected, remove the left-hand side panel (see Routine Maintenance) and trace back the leads from the unit to the connectors near the footboard. Disconnect the leads and measure the resistance between them using a multimeter. If the internal resistance is about 10 ohms, the unit is probably serviceable, but if infinite resistance is shown it can be assumed to be faulty and should be renewed. Note that the machine should have stood for at least 10 minutes before the test is made.

4 Further testing of the unit requires the removal of the carburettor and some method of applying compressed air to the richening circuit, via the air intake passage. This is located on the intake side of the carburettor throat, near the top. A foot or hand pump can be used, provided a suitable nozzle can be arranged to suit the passage.

5 Let the carburettor stand for at least 30 minutes, then apply air to the richening circuit. If it appears blocked, the cold start device is probably faulty. If air passes freely, connect the machine's battery to the cold start device leads and wait five minutes. If there is no resistance to the applied air pressure, renew the cold start device.

6 If the above checks are difficult to carry out at home, take the machine to a Honda dealer, who will be able to advise on the condition of the unit. Note that the sealed construction rules out any attempt at repair. The unit is held in place by two cross-head screws. Before refitting the cold start device, check the valve for signs of wear or damage, and renew the O-ring if it is damaged or broken.

6.6a Automatic choke unit is retained by plate and two screws

6.6b Check condition of O-ring (arrowed) and renew as required

7 Carburettor: removal and refitting

1   To gain access to the carburettor, remove the left-hand side panel as described in Routine Maintenance. Remove the two bolts which retain the air filter to the top of the transmission casing and lift it away. Trace and disconnect the leads from the cold start device to their connectors near the footboard and separate them. Unscrew the carburettor top and withdraw the throttle valve assembly. This need not be dismantled unless it requires specific attention, and can be lodged on the frame until the carburettor is refitted.

2   Free the fuel pipe from the carburettor by squeezing together the 'ears' of the wire clip and sliding it away from the stub. Push the pipe off the stub using a small screwdriver, rather than attempting to pull it off; this will only cause it to grip the stub more firmly. Unscrew the two mounting bolts and lift the carburettor away from the inlet adaptor. Take care not to lose the heat insulator which will probably drop free as the carburettor is removed. Plug the inlet adaptor with clean cloth to prevent dirt from entering the inlet port.

3   The carburettor can be installed by reversing the removal sequence after removing the plug from the inlet adaptor. Make sure that the O-ring is intact and in position, and do not omit the heat insulator spacer. Tighten the mounting bolts to 0.9 – 1.2 kgf m (7 – 9 lbf ft) only. If the throttle valve was removed from the cable, fit the dust seal and carburettor top to the cable, then place the spring over the cable and compress it against the top. With the needle and retainer in position in the valve, place the cable into the top of the valve with the nipple projection through the slot in its side. Slide the nipple down and into the hole in the underside of the valve, and then allow the spring to seat inside the valve.

4   Offer up the valve, ensuring that the slot engages over the guide pin and that the needle enters the needle jet. Screw down the carburettor top. Reconnect the cold start device leads, and fit the air filter casing, making sure that the carburettor engages correctly in it. Check the throttle cable free play and idle speed adjustment as described later in this Chapter.

8 Carburettor: examination and renovation

1   Before dismantling the carburettor, cover an area of the work surface with clean paper or rag. This will not only prevent any components that are placed upon it from becoming contaminated with dirt, moisture or grit but, by making them more visible, will also prevent the many small components removed from the carburettor body from becoming lost.

2   Proceed to dismantle the carburettor by holding it over a small

clean container and then removing the drain screw from the base of the float chamber. Any fuel contained in the float chamber should be allowed to drain into the container. Take care to observe the necessary fire precautions whilst doing this and during the various cleaning procedures listed in the following paragraphs of this Section.

3   Unscrew the two screws which retain the float chamber to the carburettor body. Note the condition of the spring washer fitted beneath the head of each of these screws and renew if found to be flattened or broken. Carefully separate the float chamber from the carburettor body whilst taking great care not to stretch or tear the seal between the two mating surfaces.

4   Remove the domed cross head screw which serves to retain the float pivot pin in its groove in the carburettor body. Detach the float, with its pivot pin and needle, from the carburettor body. The float needle should now be displaced from the float and put aside in a safe place for examination at a later stage. The needle is very small and easily lost.

5   Locate the pilot air screw and screw it inwards until it is felt to come into contact with its seat. Count the number of turns required to do this and take care not to overtighten the screw. Make a note of the screw setting and then remove it. Carry out a similar procedure to remove the throttle stop screw. Failure to note the settings of these screws will make it less easy to 'retune' the carburettor after it has been reassembled and refitted to the machine.

6   Carefully remove the O-ring from the flange of the carburettor body and the seal from its retaining groove in the float chamber. Closely inspect each item for signs of damage and deterioration and renew each one as necessary. Note that it is considered good practice to renew both of these seals as a matter of course when the carburettor is being renovated. The main jet is situated in the centre pillar of the carburettor. Use a close-fitting screwdriver to unscrew the jet otherwise the slot in the soft jet material may be damaged. With the carburettor returned to its upright position the needle jet may fall out. It may be necessary to push the jet out from above using a soft wooden rod. The vent tubes connected to the float chamber need not be removed unless they are seen to be perished or in any way damaged and therefore require renewal.

7   Prior to examination of the carburettor component parts, clean each part thoroughly in clean petrol before placing it on a piece of clean rag or paper. Use a soft nylon-bristled brush to remove any stubborn contamination from the castings and blow dry each part with a jet of compressed air. Avoid using a piece of rag for cleaning since there is always risk of particles of lint obstructing the airways or jet orifices. Never use a piece of wire or any pointed metal object to clear a blocked jet, it is only too easy to enlarge a jet under these circumstances and increase the rate of petrol consumption. If an air line is not available, a blast of air from a tyre pump will usually suffice. If all else fails to clear a blocked jet, remove a bristle from the soft-bristled brush and carefully pass it through the jet to clear the blockage.

8   Check each casting for cracks or damage and check that each mating surface is flat by laying a straight-edge along its length. A distorted casting must be replaced with a serviceable item.

9   Ensure that the O-ring fitted to the flange of the carburettor body and the seal fitted to the float chamber are both correctly located in their retaining grooves. Examine and, if necessary, renew the small O-ring fitted to the drain screw. The springs fitted to the throttle stop and pilot air screws should now be carefully inspected for signs of fatigue and corrosion and renewed if necessary.

10 The seating area of the float needle will wear after lengthy service and should be closely examined with a magnifying glass. Wear usually takes the form of a ridge or groove which will cause the float needle to seat imperfectly. If the needle has to be renewed, remember that the needle seat will have worn in unison and in extreme cases, will also need to be renewed. Note that the needle seat forms part of the carburettor body which means that if the seat is defective then the complete carburettor body will have to be renewed. Check also that the small pin which protrudes from the end of the needle is free to move and is returned to its extended position by the action of the spring fitted beneath it. The correct action of this pin is essential in cushioning the movement of the float against the needle which in turn acts to reduce the amount of wear on the seating area of the needle.

11 Closely examine the float for signs of damage or leakage. Leakage of the float will be obvious on inspection, because the float material is translucent. A defective float must be renewed.

12 Move to the machine and inspect the throttle valve for wear. This wear will be denoted by polished areas on the external diameter.

Excessive wear will allow air to leak past the valve, thereby weakening the fuel/air mixture and producing erratic slow running. Many mysterious carburation maladies may be attributed to this defect, the only cure being to renew the valve, and if worn badly in corresponding areas, the carburettor body.

13 Examine the jet needle for scratches or wear along its length. If either the valve or needle is seen to be defective, then detach the valve from the throttle cable by grasping the valve firmly in one hand whilst compressing the return spring against the carburettor top with the other. Disengage the throttle cable from its retaining slot in the valve and place the valve, together with its return spring, on a clean work surface.

14 The jet needle may be detached from the throttle valve after removal of its retaining clip. This clip should be eased from position with either the flat of a small screwdriver or a pair of long-nosed pliers. Note the fitted position of the clip attached to the jet needle and then carefully push it from its retaining groove. Check that the needle is not bent by rolling it on a flat surface, such as a sheet of plate glass. If in doubt as to the condition of the needle, or the jet through which it passes, return it to an official Honda dealer who will be able to give further advice and, if necessary, provide a new component. Refit the jet needle clip in its original position.

15 Inspect the valve return spring for signs of fatigue, failure or severe

corrosion and renew it if found necessary. The sealing ring located within the carburettor top must be renewed if seen to be perished or in any way damaged. The procedure adopted for reassembly of the throttle valve component parts should be a direct reversal of that used for dismantling.

16 Prior to reassembly of the carburettor, check that all the component parts, both new and old, are clean and laid out on a piece of clean rag or paper in a logical order. On no account use excessive force when reassembling the carburettor because it is easy to shear any one of the screw threads. Furthermore, the carburettor is cast in a zinc based alloy which itself does not have a high tensile strength. If any of the castings are damaged during reassembly, they will almost certainly have to be renewed.

17 Reassembly is basically a reversal of the dismantling procedure, whilst noting the following points. If in doubt as to the correct fitted position of a component part, refer either to the figure accompanying this text or the appropriate photograph.

18 When fitting the throttle stop and pilot air screws, ensure that each screw is first screwed fully in, until it seats lightly, and then set to its previously noted position. Alternatively, set the pilot air screw $2\frac{1}{8}$ turns out from fully in, the setting of the throttle stop screw will then have to be determined by following the adjustment procedure listed in Section 9 of this Chapter.

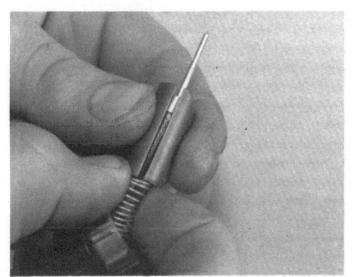

7.3 Slide throttle cable into position via slot in side of carburettor throttle valve

8.4a Float assembly is retained by a single screw (arrowed)

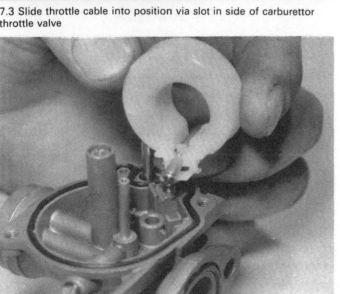

8.4b Lift away float, together with needle and pivot pin

8.7a Main jet unscrews from projection at centre of carburettor body ...

8.7b ... and is followed by the needle jet

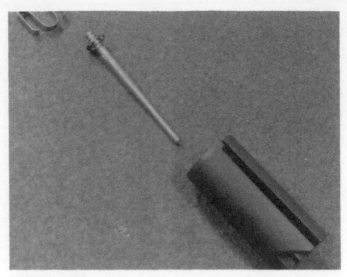

8.14 The needle and retaining clip removed from throttle valve

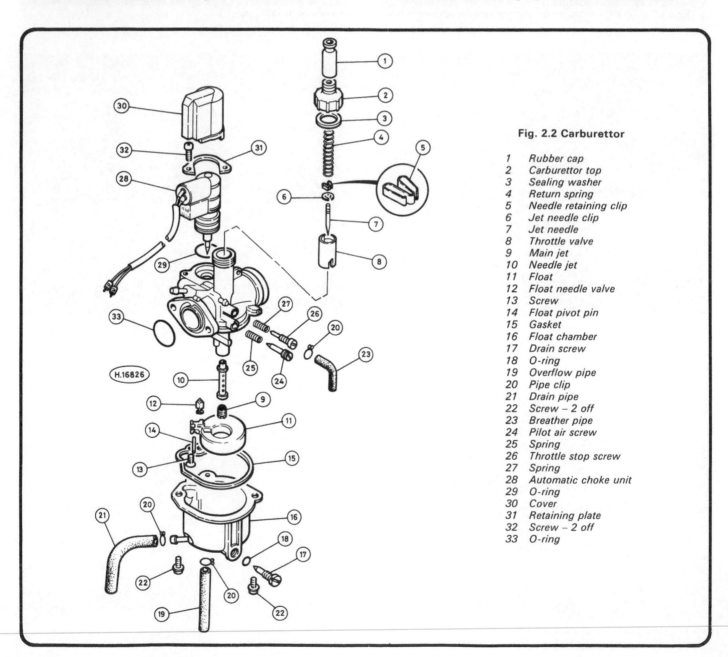

Fig. 2.2 Carburettor

1   Rubber cap
2   Carburettor top
3   Sealing washer
4   Return spring
5   Needle retaining clip
6   Jet needle clip
7   Jet needle
8   Throttle valve
9   Main jet
10 Needle jet
11 Float
12 Float needle valve
13 Screw
14 Float pivot pin
15 Gasket
16 Float chamber
17 Drain screw
18 O-ring
19 Overflow pipe
20 Pipe clip
21 Drain pipe
22 Screw – 2 off
23 Breather pipe
24 Pilot air screw
25 Spring
26 Throttle stop screw
27 Spring
28 Automatic choke unit
29 O-ring
30 Cover
31 Retaining plate
32 Screw – 2 off
33 O-ring

H.16826

9 Carburettor: adjustment

1 The float level should be checked whenever significant running problems are experienced, although it should be noted that if these occur suddenly, they may be due to contamination of the fuel and the resulting jet blockages; check this first, having removed and stripped the carburettor as described in Section 8, then carry out the float level check before refitting the float bowl.

2 Place the carburettor on a flat surface so that the float hangs vertically and the fuel valve is just closed. With the float bowl O-ring removed, the distance between the lower edge of the float and the carburettor body should be 12.2 mm (0.48 in). Honda recommend that the float be renewed if the float level is other than specified, but it is worth trying a little judicious bending of the small tab which controls the valve before buying a new one. Do not try to bend it too far; it is easily broken off. Note also that wear in the float valve and seat will affect the float level, and this should be checked. Once the float level is set correctly, reassemble and refit the carburettor.

3 The engine idle speed must be set correctly or problems will arise; if too high the machine will tend to 'creep' when at a standstill, whilst if it is too low the engine may stall repeatedly. The idle speed should be checked whenever the carburettor has been dismantled or the settings disturbed, and at the recommended service interval. Refer to Routine Maintenance, Six monthly checks, for details of adjustment.

4 Once the idle speed has been set correctly, check the pilot air screw setting as follows. Set the screw to the specified $2^1/8$ turns out and start the engine. Allow it to reach normal operating temperature, preferably by riding the machine for several miles. If the engine runs erratically or tends to misfire, try turning the screw in or out by up to $1/4$ turn in either direction. Find the setting which gives the fastest and most reliable tickover. Once set, check and if necessary adjust the idle speed once more.

10 Carburettor: settings

1 The various jet sizes, the throttle valve cutaway and the jet needle profile and clip position are predetermined by the manufacturer and should not warrant any change during the life of the machine. In almost every case, a running fault can be attributed to some external

cause, often a choked or damaged air filter element or exhaust system, or a blocked jet or drilling. Always check these areas first.

2 The pilot jet is integral with the carburettor body. If cleaning and checking of the fuel system components does not rectify the fault, try fitting a new spark plug to eliminate that as a cause of the problem, then carry out the idle speed and pilot screw adjustment procedure. If all else fails, renew the carburettor. It is recommended that a Honda dealer is consulted before taking this step.

11 Reed valve: examination and renovation

1 To gain access to the reed valve it will first be necessary to remove the carburettor as described earlier in this Chapter. With the carburettor removed, release the four bolts holding the exhaust system to the engine/transmission unit and lift it away. Next, remove the bolts holding the cylinder cowling and lift this clear to gain access to the inlet adaptor bolts.

2 Remove the inlet adaptor bolts and lift it away. The reed valve unit can now be removed by carefully prising it out of the inlet tract. Place some rag in the inlet tract to exclude dirt. Examine the reed valve case for wear or damage, paying particular attention to signs of deterioration around the valve seat area. Check the valve petals for cracking and look for gaps between the petal and seat, indicating weakness or wear in the petals. If any fault is found, the valve assembly must be renewed. Honda specifically warn against dismantling the valve or attempting to bend the valve stopper plate. When refitting the unit, remember to remove the rag from the inlet tract. Tighten the intake adaptor bolts to 0.8 – 1.2 kgf m (6 – 9 lbf ft).

12 Air filter element: removal and cleaning

This operation should be carried out without fail at the intervals specified in Routine Maintenance, and more frequently in particularly dusty conditions. If the element is allowed to become clogged with dust, engine performance and fuel consumption will suffer, and if the element is damaged or missing, rapid internal wear can be expected. Refer to Routine Maintenance, Six Monthly checks, for details of removal and cleaning.

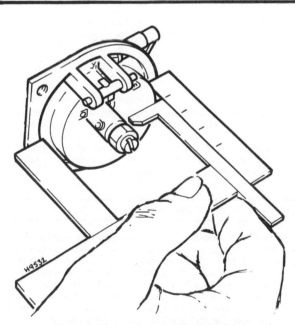

Fig. 2.3 Measuring the float height

11.2 Reed valve should be checked for wear or deterioration

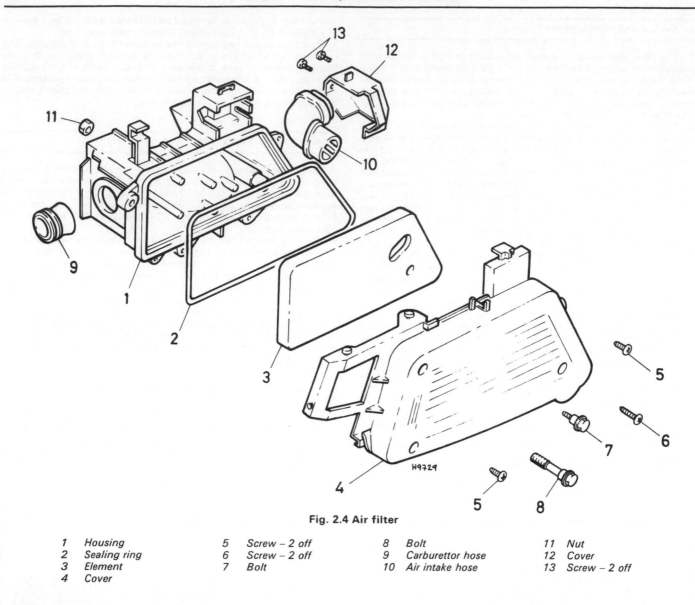

Fig. 2.4 Air filter

1	Housing	5	Screw – 2 off	8	Bolt	11	Nut
2	Sealing ring	6	Screw – 2 off	9	Carburettor hose	12	Cover
3	Element	7	Bolt	10	Air intake hose	13	Screw – 2 off
4	Cover						

13 Exhaust system: removal and refitting

1 To gain access to the exhaust system it is preferable to remove the right-hand side panel. This is described in Routine Maintenance. Remove the two nuts which retain the exhaust pipe mounting flange to the cylinder barrel, and the two bolts which secure the silencer to the edge of the crankcase. The complete assembly can now be swung clear of the engine and removed.

2 The exhaust system is refitted by reversing the removal sequence. Note that a new sealing ring should be fitted at the exhaust port, and that both the nuts and bolts should initially be fitted finger-tight. Tighten the exhaust port nuts first, followed by the silencer bolts.

14 Exhaust system: decarbonising and refinishing

1 The exhaust system is a one-piece welded unit comprising an exhaust pipe and silencer. It should be detached for cleaning if it is evident that restriction is causing a drop in engine performance. This condition is often displayed as an inability of the engine to reach maximum revolutions. Refer to Routine Maintenance, Yearly heading, for details of decarbonisation.

2 When new, the system is protected by a high temperature paint finish which prevents rusting, but in time this will begin to flake off

leaving the silencer unprotected. If ignored, the thin sheet steel silencer box will quickly rust through, necessitating a new replacement item. This can be avoided by refinishing the assembly every six months or so. The silencer and pipe can be cleaned of rust by using a wire brush and then degreased before having a fresh coat of heat resistant paint applied to their surfaces. One of the very high temperature (VHT) coatings such as Sperex is recommended.

15 Oil tank: removal and refitting

1 Remove the rack and both side panels as detailed in Routine Maintenance. Trace and disconnect the oil level sensor leads and remove the oil filler cap. Disconnect the oil feed pipe from the tank at the pump end and allow the oil to drain into a suitable container. The drained oil can be kept for reuse unless it has become contaminated with dirt, water or oil of the wrong type. The tank is now held by a single bolt and the filler neck, which locates in a hole through the battery tray top. Remove the bolt and lift the tank away.

2 While the tank is removed, note that it is a convenient time to check the operation of the oil level sensor (see Chapter 6) and to clean the tank and its outlet filter as described in Section 16. The tank can be refitted by reversing the removal operation, but do not fit the side panels and rack until the tank has been filled and the oil lines bled of air as described in Section 18.

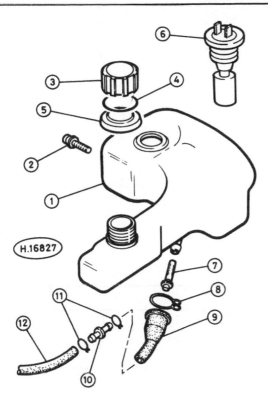

Fig 2.5 Oil tank

1	Oil tank	7	Filter
2	Bolt	8	Clamp
3	Filler cap	9	Adaptor
4	Gasket	10	Union
5	Seal	11	Clip – 2 off
6	Oil level switch	12	Oil pipe to pump

16 Oil tank filter: removal and cleaning

1   The tank filter can be removed for cleaning with the tank in place, although it is easier to carry out this job with the tank removed. Removal of the tank also permits thorough cleaning of the tank to remove any residual debris. Start by removing the side panels as described in Routine Maintenance. Drain the oil tank as described in the previous Section, and remove the tank from the frame. Squeeze together the ears of the clip which secures the filter adaptor to the tank outlet stub, then slide the assembly off the tank.

2   If oil starvation problems have been experienced, yet the filter element is seen to be clean then the oil feed pipe must be checked for blockage or signs of damage leading to leakage of oil. Renew the pipe, if necessary, and proceed to clean the filter element as follows.

3   The element should be cleaned by directing a jet of compressed air through its open end. Do not attempt to direct air through the sides of the element as this will only serve to collapse the fine filter mesh, making renewal of the element necessary. Take care to wear adequate eye protection against any spray back of oil from the element and remember to observe the necessary fire precautions. If necessary, any stubborn traces of contamination can be removed by gentle rubbing of the element with a soft-bristled brush; a used toothbrush is ideal. On completion of cleaning, closely inspect the element for any splits or holes that will allow the passage of sediment through it and into the pump. Renew the element if it is in any way defective.

4   Fitting of the filter components and oil feed pipe is a direct reversal of the removal procedure, whilst noting the following points. After having refilled the oil tank with Honda 2-stroke injector oil, or equivalent, prime the oil feed pipe and bleed the oil pump of air in accordance with the instructions given in Section 18 of this Chapter. Do not refit the side panels until the machine has been run and a comprehensive check for oil leaks carried out at all disturbed connections.

17 Oil pump: removal and refitting

1   The oil pump is mounted at the front of the engine/transmission unit and can be reached after the side panels have been removed (see Routine Maintenance). Disconnect and plug the pipe from the oil tank, then disconnect the delivery pipe at the inlet adaptor. The pump is held in place by a retainer plate which is secured by one of the cylinder cowling bolts. Once the retainer has been removed, twist and pull the pump out of the crankcase.

2   If the pump has failed internally (check this by attempting to bleed the system) little can be done other than to renew the unit. Similarly, a new pump must be fitted if there is obvious damage to the body or driving gear, or signs of leakage.

3   Before refitting the pump, fit a new O-ring into the groove in the mounting hole. Grease the pump body, then slide it into place. Refit the retainer, then reconnect the oil tank and delivery pipes after priming them with oil. Do not omit to bleed the system as described in the next section.

18 Oil pump and pipes: bleeding

1   The lubrication system must be bled to remove all traces of air whenever the tank has been allowed to run dry or if any of the pipes or the pump has been disturbed. Start by disconnecting the pipe from the tank at its lower end and allowing the oil to drain into a suitable receptacle. Wait until all traces of air have been expelled, then reconnect the pipe to the pump stub. Check and top up the oil tank to the maximum level using Honda 2-stroke injector oil, or equivalent.

2   Remove the oil delivery pipe between the oil pump and the inlet adaptor and fill it completely with oil. Taking care not to spill the oil, reconnect the delivery pipe at the pump end. Start the engine, and check that oil is forced from the delivery pipe within one minute of starting. If oil fails to emerge, repeat the priming stages and try again. If oil still fails to emerge after repeated attempts it will have to be assumed that the pump has failed internally, necessitating its renewal.

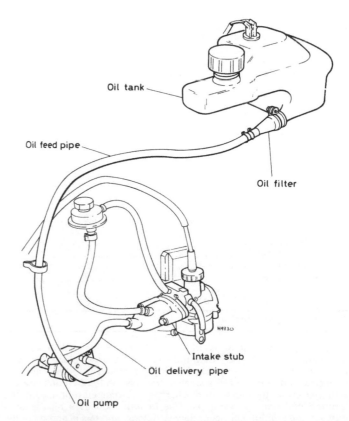

Fig. 2.6 Correct routing of the engine oil pipes

Chapter 3 Ignition system

Refer to Chapter 7 for information relating to the SA50 Met-in

Contents

General description 1
Checking the ignition system: general information 2
Spark plug: maintenance and renewal 3
Checking spark plug operation 4
Checking the ignition system: tracing wiring faults 5
Ignition source coil: testing 6

Ignition pickup coil: testing 7
CDI unit: testing 8
Ignition HT coil: testing 9
Ignition HT lead: checking 10
Ignition timing: checking 11

Specifications

Ignition system

Type .... CDI (capacitor discharge ignition)

Ignition coil

Primary winding resistance .... 0.2 – 0.3 ohms
Secondary winding resistance:
    With plug cap .... 8.2 – 9.3 K ohms
    Without plug cap .... 3.4 – 4.2 K ohms

Ignition pickup

Resistance .... 50 – 200 ohms

Ignition source coil

Resistance.. 0.6 – 1.0 K ohms

Ignition timing

Advance .... 13° BTDC @ 1800 ± 100 rpm (fixed)

Spark plug

	NGK	ND
Make	NGK	ND
Type:		
Standard	BPR6HSA, BPR6HS	W20FPR-L, W20FPR
Cold climates	BPR4HSA, BPR4HS	W14FPR-L
Continuous high speed	BPR8HSA, BPR8HS	W24FPR-L, W24FPR
Electrode gap	0.6 – 0.7 mm (0.024 – 0.028 in)	

1 General description

The ignition system is of the fully electronic CDI (capacitor discharge ignition) type. When the engine is running, electrical power is fed to the CDI unit, where a charge is stored in a capacitor, the capacitor acting rather like a small battery. When the CDI unit receives a trigger pulse, the capacitor discharges its stored current through the ignition coil primary windings. This in turn induces a high tension pulse in the ignition coil secondary windings, and this is applied to the spark plug. The high tension pulse jumps across the plug electrodes to earth, the resulting spark igniting the fuel/air mixture in the cylinder.

The trigger pulse is provided by a small magnetic pickup incorporated in the flywheel generator; a magnet in the rotor sweeps past a small pickup coil, generating a tiny signal voltage. This is sufficient to trigger the CDI unit, discharging the capacitor. The absence of mechancial parts in the system means that no maintenance is required, and in fact no adjustment is possible. This type of ignition system is well established and can be expected to be reliable over a long period of service. Where faults do occur, they tend towards failure of a single component in the system, and can be resolved by renewing the affected component.

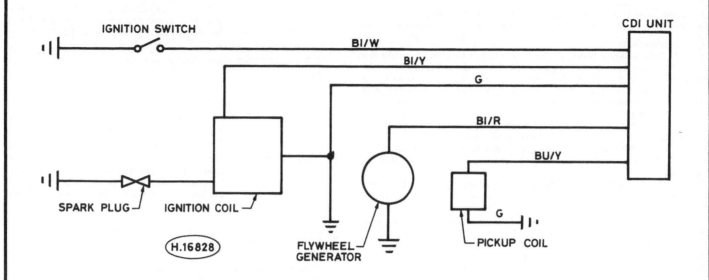

Fig. 3.1 Ignition system circuit diagram

BI/R	Black and red	BI/Y	Black and yellow
BI/W	Black and white	Bu/Y	Blue and yellow

G   Green

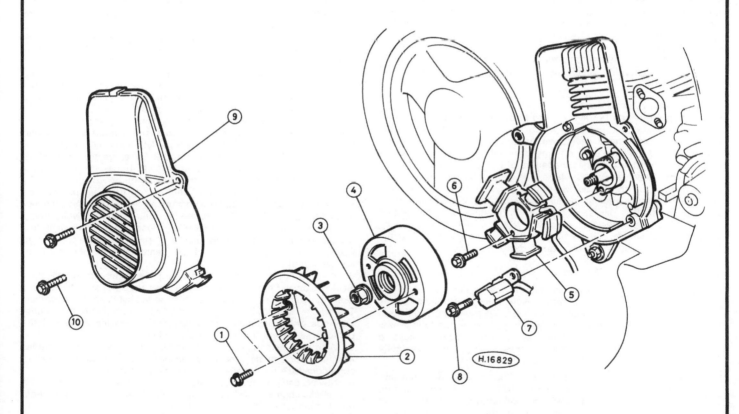

Fig. 3.2 Flywheel generator and fan assembly

1	Bolt – 2 off	5	Stator	8	Bolt
2	Fan	6	Bolt – 2 off	9	Fan cowling
3	Nut	7	Pickup coil	10	Bolt – 2 off
4	Rotor				

2 Checking the ignition system: general information

Test equipment

1 It is particularly important that any attempt to check a suspected fault in the ignition system is approached in a logical and methodical fashion. A haphazard attempt at resolving the fault is likely to prove time consuming, confusing and often inconclusive. The range of tests which can be performed at home is distinctly limited, many of the more detailed tests requiring specialised test equipment unavailable to the owner by virtue of its cost. Whilst this means that in some cases it will be necessary to resort to a Honda dealer for testing work or for confirmation of a suspected fault, it is quite reasonable for the owner to establish the general nature of the problem and to effect a cure in most instances.

2 A certain amount of specialist equipment will be needed to carry out the tests described in this Chapter. For example, a stroboscopic timing lamp is needed to check the ignition timing, and one of two specified test meters will be required to make accurate CDI unit resistance checks. For a full test of the CDI unit, a special test rig is necessary, and this will not be available for home use. In addition to the above, an extractor (Part Number 07733-0010000) will be needed to draw off the generator rotor if access to the generator stator is necessary.

3 Of the above, the rotor extractor is well worth purchasing, and a stroboscopic timing lamp should be considered in view of its relatively low cost and general usefulness. The Sanwa and Kowa test meters specified by Honda are probably not worthwhile purchases for most owners, whilst the CDI unit tester is most definitely out of the question for home use. This effectively limits the owner to preliminary diagnosis of the fault, which can then be confirmed by a Honda dealer, either by testing or substitution. This may seem a somewhat pessimistic situation for the enthusiastic owner, but it should be stressed that the majority of the more common faults are fairly easily resolved before detailed testing becomes necessary. A multimeter capable of reading in ohms and kilo ohms may be used to check all the ignition components except the CDI unit, where only the Kowa or Sanwa tester should be used. Simple continuity checks can be made by using a battery and bulb test circuit.

Defining the type of fault

4 Ignition faults can be divided into two main categories; partial or intermittent failure and complete failure. In the case of the former, the problem will often have developed gradually, commencing perhaps with unreliable starting, poor running or misfiring. Complete failure is less usual, and often indicates the failure of one of the ignition system components. Having decided which type of fault has occurred, follow the sequence of checks as listed below, referring to the relevant Sections of this Chapter for details. Note that the checks are similar, whether the fault is partial or total. When dealing with the ignition system, treat it with respect, especially when switched on. A shock from the CDI system can be unpleasant or even dangerous to the owner, and may also cause damage to the CDI unit itself.

Check	Details
a) Spark plug	Always check the spark plug first; it is the single most likely cause of failure simply because it will eventually wear out, unlike the rest of the system. It is preferable to fit a new spark plug as a precautionary measure because sometimes a plug can prove faulty while appearing quite normal.
b) Spark	Fit a new spark plug into the plug cap and arrange it so that the body of the plug is in firm contact with a good earthing point, such as an unpainted part of the engine/transmission unit. Switch on the ignition and operate the kickstart lever while watching the spark plug electrodes. In a normal system, a fat, regular bluish spark should be seen. A weak, irregular or yellow spark or no spark at all indicates a fault somewhere in the system.
c) Wiring and connectors	Using the wiring diagram at the back of this manual, check through all of the ignition wiring, looking for damaged or broken leads.

Check all connectors for looseness or corrosion. Remember that earth connections are as important as the wiring connections.

d)	Ignition coil, HT lead and plug cap	Check the ignition coil connections as described above. Test the coil resistances as described later in Section 9.
e)	Pickup coil	Check the pickup coil resistance as described in Section 7. Renew coil if defective.
f)	CDI unit	Check the CDI unit resistances as described in Section 8. If fault persists, have the system checked by a Honda dealer.
g)	Ignition timing	The ignition timing should be checked in cases where an intermittent fault or poor running cannot be traced to another cause, see Section 11. Inaccurate timing means renewal of the CDI unit.

3 Spark plug: maintenance and renewal

1 Honda recommend the use of NGK or ND spark plugs, the correct grades being shown in the Specifications Section of this Chapter. It will be noted that plug grades are given for standard, cold-weather and constant high speed conditions. In normal circumstances the standard grade should be used, but if the machine is used in low temperatures, or is ridden hard in hot conditions, a change of grade may be required. If problems are experienced it is advisable to consult a Honda dealer for advice on any proposed change. When fitting a new plug, always set the electrode gap to the prescribed 0.6 – 0.7 mm (0.024 – 0.028 in) as described below. It is quite permissible to clean and readjust a used plug, which can then be reused. Many owners, however, may prefer to renew the plug in preference to cleaning. Plugs are not expensive, and this choice of action removes any risk of being let down by a 'tired' plug at a later date. If cleaning and refitting is preferred, this is described below.

2 The plug should be cleaned thoroughly by using one of the following methods. The most efficient method of cleaning the electrodes is by using a grit blasting machine. It is quite possible that a local garage or motorcycle dealer has one of these machines installed on the premises and will be willing to clean any plugs for a nominal fee. Remember, before fitting a plug cleaned by this method, to ensure that there is none of the blasting medium left impacted between the porcelain insulator and the plug body. An alternative method of cleaning the plug electrodes is to use a small brass-wire brush. Most motorcycle dealers sell such brushes which are designed specifically for this purpose. Any stubborn deposits of hard carbon may be removed by judicious scraping with a pocket knife. Take great care not to chip the porcelain insulator round the centre electrode whilst doing this. Ensure that the electrode faces are clean by passing a small fine file between them; alternatively, use emery paper but make sure that all traces of the abrasive material are removed from the plug on completion of cleaning.

3 To reset the gap between the plug electrodes, bend the outer electrode away from or closer to the central electrode and check that a feeler gauge of the correct size can be inserted between the electrodes. The gauge should be a light sliding fit. Never bend the central electrode or the insulator will crack, causing engine damage if the particles fall in whilst the engine is running.

4 With some experience, the condition of the spark plug electrodes and insulator can be used as a reliable guide to engine operating conditions. See the accompanying colour photographs.

5 Always carry a spare spark plug of the correct type. The plug in a two-stroke engine leads a particularly hard life and is liable to fail more readily than when fitted to a four-stroke.

6 Beware of overtightening the spark plug, otherwise there is risk of stripping the threads from the aluminium alloy cylinder head. The plug should be sufficiently tight to seat firmly on its sealing washer, and no more. Use a spanner which is a good fit to prevent the spanner from slipping and breaking the insulator.

7 If the threads in the cylinder head strip as a result of overtightening the spark plug, it is possible to reclaim the head by the use of a Helicoil thread insert. This is a cheap and convenient method of replacing the threads; most motorcycle dealers operate a service of this nature at an economic price.

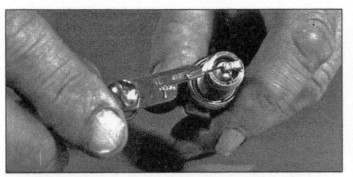

Electrode gap check - use a wire type gauge for best results

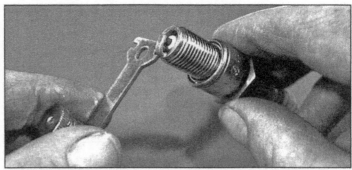

Electrode gap adjustment - bend the side electrode using the correct tool

Normal condition - A brown, tan or grey firing end indicates that the engine is in good condition and that the plug type is correct

Ash deposits - Light brown deposits encrusted on the electrodes and insulator, leading to misfire and hesitation. Caused by excessive amounts of oil in the combustion chamber or poor quality fuel/oil

Carbon fouling - Dry, black sooty deposits leading to misfire and weak spark. Caused by an over-rich fuel/air mixture, faulty choke operation or blocked air filter

Oil fouling - Wet oily deposits leading to misfire and weak spark. Caused by oil leakage past piston rings or valve guides (4-stroke engine), or excess lubricant (2-stroke engine)

Overheating - A blistered white insulator and glazed electrodes. Caused by ignition system fault, incorrect fuel, or cooling system fault

Worn plug - Worn electrodes will cause poor starting in damp or cold weather and will also waste fuel

8   Before fitting the spark plug in the cylinder head, coat its threads sparingly with a graphited grease. This will prevent the plug from becoming seized in the head and therefore aid future removal.
9   When reconnecting the suppressor cap to the plug, make sure that the cap is a good, firm fit and is in good condition; renew its rubber seals if they are in any way damaged or perished. The cap contains the suppressor that eliminates both radio and TV interference.

4   Checking spark plug operation

1   This is an essential first step in tracing any suspected ignition fault. With the ignition switched off, remove the spark plug from the cylinder head and refit it in the plug cap. Place the metal body of the plug in firm contact with a good earth point, such as the unpainted metal of the engine/transmission unit, and in a position where the plug electrodes can be viewed while the kickstart is operated.
2   Switch on the ignition and operate the kickstart lever while watching the plug electrodes. If all is well, a regular, fat blueish spark should be seen jumping across the electrode gap. A thin, yellowish or irregular spark is indicative of a fault. If the spark looks weak, try fitting a new plug, having set the electrode gap correctly. If this fails to improve the spark, check the ignition coil resistances as described in Section 9.
3   Where no spark at all is evident, check for a break in the ignition system wiring or for a fault in the ignition switch. Try turning the switch on and off a few times if it is suspected that the switch contacts are dirty or corroded. If the fault persists it will be necessary to check through the ignition system components as described in the following Sections until the cause of the problem is isolated.

5   Checking the ignition system: tracing wiring faults

1   As mentioned, the single most likely cause of ignition failure is a damaged or broken wiring connection. It is assumed that the spark plug has been removed and checked by substitution to eliminate it as the source of the problem. Where possible, a multimeter set on the resistance position should be used to establish the likely location of the faulty connection. Failing this, connect up a dry battery and bulb as shown in the figure accompanying this text. The two probe leads can be fitted with small crocodile clips and can be connected to either end of a circuit to indicate continuity by lighting the bulb. This latter method will work just as well as a multimeter in this application, but whichever method is chosen, it is recommended that the machine's electrical system is isolated by disconnecting the battery (negative lead first).
2   The wiring should be checked in conjunction with the wiring diagram at the back of this Manual. Work in a logical sequence and include the ignition switch in the checks. Full details of checking this switch are included in Chapter 6 of this Manual.
3   Prior to testing each wiring run, visually examine all connections and terminals for signs of corrosion or arcing, cleaning or renewing each one as required. Check also that no wiring has been trapped between, or is fraying against, any moving frame or engine components. If no faults are discovered in the wiring runs and connections or in the ignition switch, then it can be assumed that the fault must lie in the CDI unit, the pickup assembly, the ignition coil or the HT lead and suppressor cap. Check also that power is being supplied to the ignition system by the ignition source coil.

6   Ignition source coil: testing

1   The ignition source coil is mounted on the flywheel generator stator plate and provides power for the system. If it is suspected that a fault lies in the source coil this can be checked by using a multimeter to test the coil resistance. Start by removing the left-hand side panel (see Routine Maintenance) to reveal the wiring connectors grouped near the footboard.
2   Trace the flywheel generator wiring back to the connectors, and

separate the black/red lead at its connector. Using a multimeter set on the resistance range, measure the resistance between the black/red lead and earth. A reading of 0.6 – 1.0 K ohm should be shown.
3   If the test shows a reading of zero ohms, the coil insulation will have broken down, allowing the coil to short to earth. If on the other hand, a reading of infinite resistance is shown, this indicates that the coil windings have broken. Either fault will require the renewal of the stator assembly, but before ordering a new item it is worth having your findings confirmed by an expert. The alternatives to renewal are to consult an auto-electrical specialist, who may be able to rewind the faulty coil, or to consider buying a secondhand stator from a motorcycle breaker. Removal and refitting of the generator is described in the relevant sections of Chapter 1.

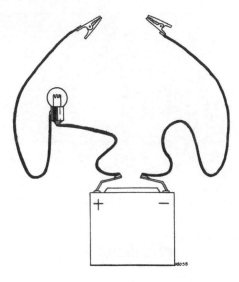

Fig. 3.3 Battery and bulb test circuit for checking the wiring

7   Ignition pickup coil: testing

1   The ignition pickup coil is mounted on the crankcase near the edge of the generator rotor and supplies the triggering impulse for the ignition system. If it is suspected that a fault lies in the pickup coil this can be checked by using a multimeter to test the coil resistance. Start by removing the left-hand side panel (see Routine Maintenance) to reveal the wiring connectors grouped near the footboard.
2   Trace the flywheel generator wiring back to the connectors, and separate the blue/yellow lead at its connector. Using a multimeter set on the resistance range, measure the resistance between the blue/yellow lead and earth. A reading of 50 – 200 ohms should be shown.
3   If the test shows a reading of zero ohms, the coil insulation will have broken down, allowing the coil to short to earth. If on the other hand, a reading of infinite resistance is shown, this indicates that the coil windings have broken. Either fault will require the renewal of the stator assembly, despite the fact that the pickup coil is not physically part of the stator.
4   Honda do not supply the pickup as a separate item, but it may be worth considering two alternatives to buying a whole new stator. An auto-electrical specialist may be able to rewind the faulty coil, or a secondhand coil or stator assembly can be bought from a motorcycle breaker. Removal and refitting of the generator is described in the relevant sections of Chapter 1.

8 CDI unit: testing

1 In the event of a suspected fault in the CDI unit, a certain amount of testing is possible at home, given access to a suitable multimeter. Honda stress that only the specified meters should be used, and that readings will not necessarily be accurate on other meters. The recommended types are either the Sanwa Electrical Tester, SP-10D (Part Number 07308-0020000) or the Kowa Electrical Tester, TH-5H. In the case of the Sanwa tester, set it to the x K ohms range. On the Kowa tester, select the x 100 ohms range.

2 To gain access to the CDI unit, remove the rack and side panels (see Routine Maintenance). The unit is mounted on the left-hand side of the machine, just below the oil tank. Trace the wiring back to the connector on the underside of the unit and unplug it. With the unit on the bench, check the various resistances as shown in the accompanying table. The CDI unit terminals are identified in the line drawing which accompanies the table.

3 It must be stressed that the above test gives only a rough indication of the condition of the unit. A full test can only be conducted using a special test rig which may be available at a local Honda dealer. Failing this, the only alternative is to check the unit by substitution.

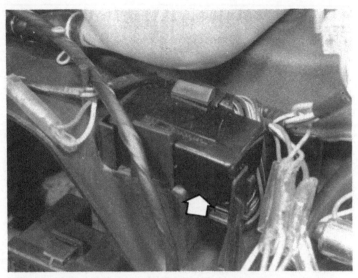

8.2 CDI unit is mounted on left-hand side of machine, just below the oil tank

RANGE SANWA: RX kΩ KOWA: RX 100Ω

(−)PROBE (+)PROBE	SW	EXT	PC	E	IGN
SW		∞	∞	∞	∞
EXT	0.1—10		∞	∞	Needle swings then returns" or ∞
PC	0.5—200	0.5—50		1—50	∞
E	0.2—30	0.1—10	∞		∞
IGN	∞	∞	∞	∞	

H.16830

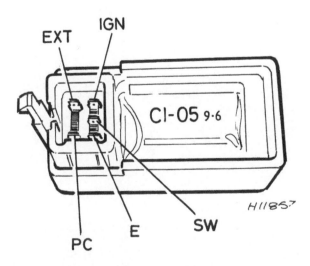

Fig. 3.4 CDI unit test

9 Ignition HT coil: testing

1 The ignition HT coil is mounted in a housing on the top of the air filter casing. It can be removed for testing after disconnecting the plug cap and low tension leads and lifting the lid of the housing. If it is suspected that a fault lies in the ignition coil this can be checked by using a multimeter to test the coil resistances. Start by removing the left-hand side panel (see Routine Maintenance) to gain access to the coil, which should be removed for testing.

2 Using a multimeter, check the resistance of the primary windings by connecting a probe to each of the low tension terminals at the back of the coil. A reading of 0.2 – 0.3 ohms should be indicated.

3 Next, measure the secondary resistance by connecting one probe to the lower of the two low tension terminals, and the other to the inside terminal of the plug cap. A reading of 8.2 – 9.3 K ohms should be shown.

4 To eliminate the plug cap as a source of trouble, remove it from the end of the plug lead and repeat the test, where a reading of 3.4 – 4.2 K ohms should be obtained. If this last test gives the correct figure, whilst the previous test showed a discrepancy, the fault can be attributed to the plug cap.

5 If either the first or last test gave incorrect resistance readings, the coil itself is suspect. Note that the resistance tests are not exhaustive; for a more accurate diagnosis, have the coil tested on a proper coil testing machine.

10 Ignition coil high tension lead: checking

1 Erratic running faults and problems with the engine suddenly cutting out in wet weather can often be attributed to leakage from the high tension lead and spark plug cap. If this fault is present, it will often be possible to see tiny sparks around the lead and cap at night. One cause of this problem is the accumulation of mud and road grime around the lead, and the first thing to check is that the lead and cap are clean. it is often possible to cure the problem by cleaning the components and sealing them with an aerosol ignition sealer, which will leave an insulating coating on both components.

2 Water dispersant sprays are also highly recommended where the system has become swamped with water. Both these products are easily obtainable at most garages and accessory shops. Occasionally, the suppressor cap or the lead itself may break down internally. If this is suspected, the components should be renewed.

3 Where the HT lead is permanently attached to the ignition coil, it is recommended that the renewal of the HT lead is entrusted to an auto-electrician who will have the expertise to solder on a new lead without damaging the coil windings.

11 Ignition timing: checking

1 The ignition timing check need only be performed if there is some reason to suspect its accuracy. No adjustment is possible, and the only reason for a timing error is a fault in the CDI unit. The test requires the use of a test tachometer and a stroboscopic timing lamp 'strobe', preferably of the xenon tube type, rather than the cheaper and less accurate neon variety. In the absence of this equipment, have the test carried out by a Honda dealer.

2 Remove the right-hand side panel (see Routine Maintenance) and remove the fan cowling to reveal the rotor edge. Identify the timing mark on the rotor, and the fixed index mark on the crankcase, above the rotor. Connect the tachometer and strobe, following the manufacturers' instructions, and direct the beam of the strobe at the fixed index mark. The timing is correct if the 'F' mark aligns with the index mark at 1800 ± 100 rpm. If the timing is incorrect, check the CDI unit and the source and pickup coils as described in previous Sections.

Chapter 4 Frame and suspension

Refer to Chapter 7 for information relating to the SA50 Met-in

Contents

General description 1
Front fork and steering head assembly: removal and refitting 2
Front suspension components: examination and renovation 3
Rear suspension pivots: examination and renovation 4
Rear suspension unit: examination and renovation 5
Frame: examination and renovation 6
Body panels: removal and refitting 7

Seat: removal and refitting 8
Steering lock and ignition switch: location and renewal 9
Stand: examination and maintenance 10
Kickstart lever: examination and renovation 11
Speedometer head: removal and refitting 12
Speedometer drive and cable: examination and maintenance 13

Specifications

Frame

Type .... Large diameter tubular front section welded to pressed steel rear spine

Front forks

Type .... Trailing link, twin coil spring suspension units
Travel .... 73 mm (2.9 in)
Spring free length .... 172.5 mm (6.79 in)
Service limit .... 167.3 mm (6.59 in)

Rear suspension

Type .... Pivoted engine/transmission unit supported by single coil spring suspension unit
Travel .... 63 mm (2.5 in)
Suspension unit .... Coil spring and hydraulic damper, non-adjustable
Spring free length .... 208.0 mm (8.19 in)
Service limit .... 201.8 mm (7.95 in)

Torque wrench settings

Component	kgf m	lbf ft
Steering stem top nut	8.0 – 12.0	58 – 87
Steering stem locking ring	0.5 – 1.3	4 – 10
Front wheel spindle nut	4.0 – 5.0	29 – 36
Fork to trailing link pivot bolt	2.7 – 3.3	20 – 24
Suspension unit:		
Upper mounting bolt	2.0 – 3.0	14 – 22
Lower mounting bolt	0.08 – 0.12	0.6 – 0.9
Lower mounting nut	1.5 – 2.0	11 – 14
Damper unit locknut	1.5 – 2.5	11 – 18
Rear wheel stub axle nut	8.0 – 10.0	58 – 72
Rear suspension unit:		
Upper mounting nut	3.0 – 4.5	22 – 33
Lower mounting bolt	2.0 – 3.0	14 – 22
Damper unit locknut	1.5 – 2.5	11 – 18

1 General description

The Honda Vision models employ a welded steel scooter-type frame comprising a large diameter tubular front section, welded to a pressed steel main section. The footboards are supported on welded pressed-steel outriggers, while the various ancillary parts are attached to the frame by numerous brackets. The frame is hidden by the extensive moulded plastic body panels.

Front suspension takes the form of a single long steering column, at the lower end of which is welded a short U-shaped fork, raked forward in relation to the steering column. From each end of the fork, a short trailing link runs rearward, the ends of which incorporate bosses for the front wheel spindle. The rear of each link is connected to the fork by way of small oil-damped telescopic suspension units.

Rear suspension is provided by allowing the entire engine/transmission unit to pivot at its forward end. The rear of the unit is supported by a single oil-damped telescopic suspension unit fitted with a multi-rate spring. Additional isolation from small bumps in the road surface is provided by rubber-mounting the engine, and by a pivoting, rubber-bushed mounting plate arrangement.

2 Front fork and steering head assembly: removal and refitting

1 The complete front fork and steering head assembly need only be removed from the machine if access to the steering head bearings or the steering column is required, or to attend to accident damage repairs. The front suspension components can be dealt with without disturbing the steering head, as described in Section 3 of this Chapter. Commence dismantling by removing the front panel and the legshield as described in Section 7 of this Chapter. Place wooden blocks under the front of the machine to raise the wheel well clear of the ground. It is necessary to allow plenty of room to manoeuvre the steering column out of the frame tube, and the machine must also be supported to prevent it from tipping forward.

2 If approached carefully, it is possible to avoid the complete dismantling of the handlebar nacelle assembly. Start by releasing the three screws which secure the rear section of the nacelle. Lift it upwards, easing the speedometer cable through the steering column until it can be detached at the upper end. If necessary, the cable should first be released at the wheel to obtain sufficient free play. The rear section of the nacelle can be lifted away after the instrument panel and switch wiring has been disconnected. In the case of the latter, this should be done at the main multi-pin connector. It is advisable to number each of the connector strips and also the corresponding position on the connector block so that it can be refitted in its correct position. Once the wiring has been freed, lift the nacelle rear section away and place it to one side.

3 It is normal practice to continue by dismantling the remaining handlebar controls and the turn signals and headlamp, so that the front section of the nacelle can be removed. This work can be avoided if care is taken. It will be necessary to disconnect the front brake cable at the wheel end so that it can be withdrawn from the steering column, but the rear brake cable and the various lamps can be left in place.

4 If the suspension components are to be dismantled as part of the steering head overhaul, it is advisable to remove the front wheel at this stage but this need not be done if the steering head bearings only are to be dealt with. If the speedometer cable is still connected at the wheel end, release the screw which secures it, then remove the front wheel spindle nut. Displace the spindle and lift the wheel clear of the fork ends.

5 Before proceeding further, obtain a large piece of cloth (an old bed sheet or a dust sheet is ideal for this purpose) and spread it on the ground beneath the steering head. The reason for this is to catch any of the steering head bearing balls as the steering column is removed. The balls are uncaged and are held in place by grease only. If they are allowed to drop onto a hard surface, a good deal of time can be wasted in scouring the darkest corners of the working area trying to locate those which bounce away.

6 To release the handlebar assembly a 32 mm socket or box spanner will be required. Slacken and remove the nut, then lift the handlebar assembly clear of the steering column and lodge it against the frame. Slacken and remove the steering stem locknut. Before proceeding further it will be advantageous to enlist the help of an assistant for the final stage in the removal sequence. Support the steering column, then remove the large hexagon-headed top race. Retrieve the steering head upper bearing balls and place them with the top race in a container for safe-keeping. Carefully lower the steering column, trying not to dislodge the lower bearing balls. Lift the column clear, then locate and retrieve the lower bearing balls. It will be helpful to know that there should be 26 balls in each race; check that all are accounted for before proceeding further.

7 The various component parts of the steering head assembly should now be washed in clean petrol, whilst observing the necessary fire precautions, and given a careful visual examination when dry. Look for pits or scuff marks in the cup and cone bearing surfaces. If these are not smooth and polished in appearance, it will be necessary to renew them. The ball bearings should be renewed as a matter of course if the cups and cones have to be renewed. Other than this, they should be rejected if marked or damaged in any way. If the bearings are in anything other than perfect condition, the steering of the machine will be adversely affected, and for this reason any slightly suspect part demands renewal to ensure that the machine is kept roadworthy.

8 The upper and lower bearing cups may be removed from the headstock by passing a long drift through the inner bore of the headstock and drifting out the defective item from the opposite end. The drift must be moved progressively around the cup to ensure the item leaves the headstock evenly and squarely. The lower cone fits over the steering stem and may be removed by levering it upwards or by using a bearing extractor of suitable type. If difficulty is experienced with either of these operations, the assembly should be entrusted to an official Honda dealer who will have the necessary equipment to effect an economical repair. When fitting new items, ensure that they are located squarely in their fitted positions. Do not strike the bearing cups or cone directly with a hammer to drift them into position but support the component solidly and use a length of steel tube of the appropriate diameter as a drift in conjunction with the hammer. If the end of this tube is cut square to its length, then it will keep the cup or cone in question square in its location.

9 If it is found necessary to remove the front mudguard from its location over the fork legs, this may be done by unscrewing the two bolts which retain it in position before lifting the mudguard over the top of the steering stem.

10 The steering head assembly is reassembled in the reverse order to that given for dismantling. Ensure that the bearings are generously lubricated with a high melting-point, lithium based grease and that all 26 balls are included in each race. The bearings are adjusted for free play by slackening or tightening the upper bearing cone. It will be found that the cone can be tightened considerably from the finger-tight position without having any undue effect upon the ease with which the handlebars can be turned. This does mean, however, that a load of several tons is unwittingly applied to the bearings, which will be rapidly destroyed. When setting the cone, it is necessary to remove all discernible play, but no more. The cone is secured in position by tightening the locking ring against it. The locking ring must be tightened to the specified torque setting of 4 – 10 lbf ft (0.5 – 1.3 kgf m).

11 As a guide to the correct adjustment of the bearing cone, only very slight pressure should be needed to start the front wheel turning to either side under its own weight when it is raised clear of the ground. Check also that the bearings are not too slack; there should be no discernible movement of the forks, in the fore and aft direction.

12 When refitting the handlebars to the steering stem, align the tab projecting from the handlebar centre bracket with the groove cut in the steering stem before lowering the handlebars into position and fitting their retaining nut. This nut must be tightened to the specified torque loading of 58 – 87 lbf ft (8 – 12 kgf m).

13 Re-route and reconnect all disturbed electrical wires and control cables. Check that none of these wires or cables are likely to become chafed against moving cycle components and ensure that each disturbed control is in correct adjustment and operates smoothly and efficiently. Use this opportunity to lubricate the various controls. Full information on fitting the front wheel and adjusting both the front and rear brakes is contained in Chapter 5 of this Manual.

14 With the handlebar nacelle and the front body panel assembly refitted to the machine, fit the headlamp unit and align the beam in accordance with the instructions given in Section 17 of Chapter 6. Carry out a final check before taking the machine on the road to ensure that all the electrical switches mounted on the handlebar assembly function correctly and operate their respective components.

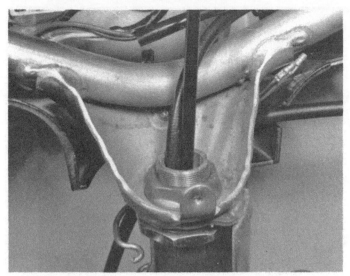

2.3 Speedometer and front brake cable are routed through the centre of the steering column

2.6a Slacken handlebar retaining nut ...

2.6b ... and lift assembly away from column. Note locating tabs and corresponding slots in column

2.6c Release the adjuster nut ...

2.6d ... and lower steering column assembly clear of frame

2.10 Steering head balls can be held in place with grease during assembly

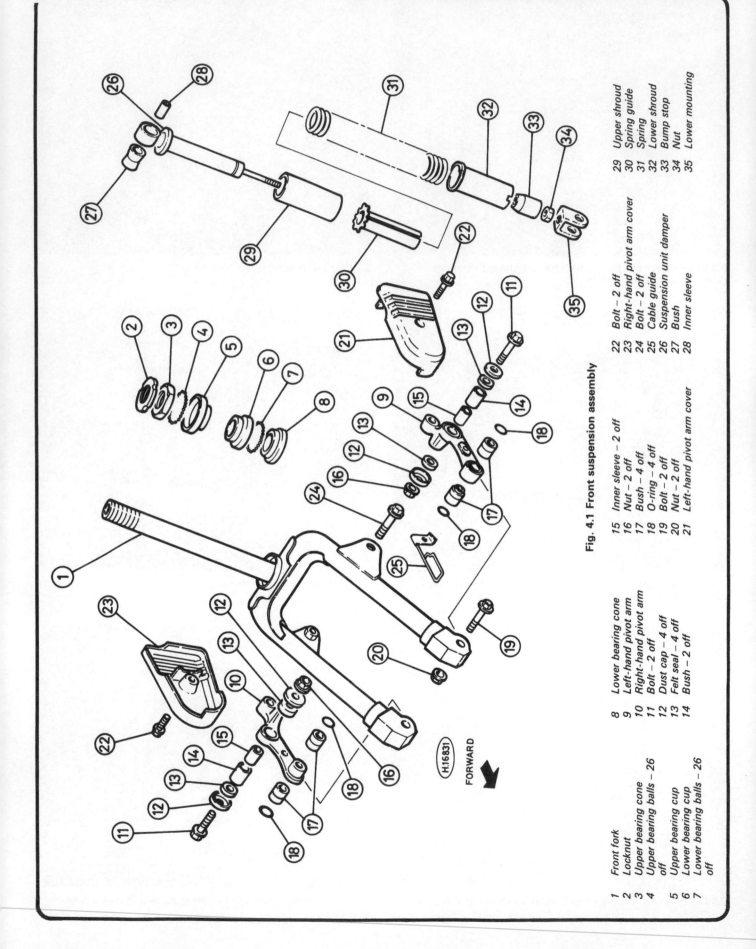

Fig. 4.1 Front suspension assembly

1	Front fork	15	Inner sleeve – 2 off
2	Locknut	16	Nut – 2 off
3	Upper bearing cone	17	Bush – 4 off
4	Upper bearing balls – 26 off	18	O-ring – 4 off
5	Upper bearing cup	19	Bolt – 2 off
6	Lower bearing cup	20	Nut – 2 off
7	Lower bearing balls – 26 off	21	Left-hand pivot arm cover
8	Lower bearing cone	22	Bolt – 2 off
9	Left-hand pivot arm	23	Right-hand pivot arm cover
10	Right-hand pivot arm	24	Bolt – 2 off
11	Bolt – 2 off	25	Cable guide
12	Dust cap – 4 off	26	Suspension unit damper
13	Felt seal – 4 off	27	Bush
14	Bush – 2 off	28	Inner sleeve
		29	Upper shroud
		30	Spring guide
		31	Spring
		32	Lower shroud
		33	Bump stop
		34	Nut
		35	Lower mounting

H.16831

FORWARD

3 Front suspension components: examination and renovation

1 As mentioned previously, access to the front suspension components does not require the removal of the steering column assembly from the frame (although the accompanying photographs show the dismantling sequence on the workbench for clarity). Start by supporting the front of the machine so that the wheel is raised clear of the ground, by placing wooden blocks beneath the footboard outrigger arms. Check that the machine is stable, then remove the front wheel. For full details covering wheel removal see Chapter 5. If the suspension units are to be detached, remove the front panel and legshield as detailed in Section 7 of this Chapter, otherwise they can be left in position.

2 Remove the single bolt which retains each of the black plastic covers to the suspension pivot arms. Lift away the covers to reveal the suspension unit lower mounting bolts and the pivot arm bolts. Release the suspension unit lower mounting locknuts, then unscrew the cross-head countersunk mounting bolts to free the lower end of the units. The pivot arms may be removed by releasing their mounting bolts.

3 When removing the various bushes and seals from the pivot arms, place them in separate containers to avoid interchanging parts from one arm to the other. Displace and remove the dust caps and felt seals at each end of the suspension unit mounting boss on each pivot arm, then displace the steel bush from the internal bore. Next, remove the large steel pivot collars from the ends of the arms. These are sealed by O-rings and may prove reluctant to come out, particularly if the bores have corroded. If difficulty is experienced, soak the assembly in penetrating fluid, then try passing a thin drift through the bore of one collar and tapping out the opposite collar.

4 Examine the various components for signs of wear or damage, renewing parts as required. Light corrosion can be removed using fine abrasive paper, provided that a significant amount of material is not removed. Clean all parts thoroughly with petrol, taking suitable precautions to avoid fire risks, and allow the petrol to evaporate before assembly commences.

5 Fit new O-rings to the pivot collars and grease the collars and their bore in the pivot arm thoroughly before pushing them home. Grease and install the suspension unit steel bush, then fit the felt seals and dust caps. Refit the pivot arms into the fork ends and reconnect the lower ends of the suspension units. Tighten the various fasteners as follows:

Component	kgf m	lbf ft
Pivot arm mounting bolts and nuts	2.7 – 3.3	20 – 24
Suspension unit lower mounting bolts	0.08 – 0.12	0.6 – 0.9
Suspension unit locknuts	1.5 – 2.0	11 – 14

6 If the suspension units are to be removed, proceed as detailed in paragraphs 1 and 2 above, but do not disturb the pivot arm mounting bolts and nuts. With the front panel and legshield removed, release the bolts which secure the front mudguard to the steering column. Lift the mudguard sufficiently to gain access to the suspension unit upper mounting bolts, taking care not to damage the paint finish. Remove the upper mounting bolts and lift the units away.

7 The units can be dismantled to some extent, but this requires the use of special tools to compress the springs safely. These tools are unlikely to be available to the owner, so it is recommended that you should entrust this work to a Honda dealer. The springs can be renewed if they have become weakened, but the internal damper unit is sealed and cannot be dismantled. For owners able to borrow a suitable compressor, note that the service limit for the spring is 167.3 mm (6.59 in). The spring should be fitted with the tightly-wound coils facing downwards. Use Loctite or similar on the locknut threads and tighten to 1.5 – 2.5 kgf m (11 – 18 lbf ft).

8 The suspension units are installed by reversing the removal sequence. Note that the upper mounting bolts should be tightened to 2.0 – 3.0 kgf m (14 – 22 lbf ft), the lower mounting bolts to 0.08 – 0.12 kgf m (0.6 – 0.9 lbf ft) and the lower mounting bolt locknuts to 1.5 – 2.0 kgf m (11 – 14 lbf ft). Lower the front mudguard into position and fit the mounting bolts, tightening them securely. Refit the legshield and front panel, and install the front wheel, remembering to adjust the front brake before the machine is ridden.

3.2a Remove single retaining bolt and lift away plastic covers

3.2b Free the suspension unit lower mounting bolts ...

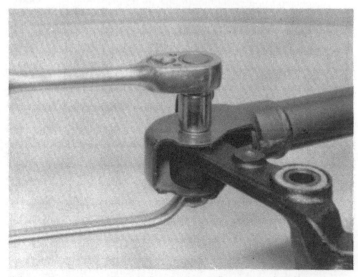

3.2c ... then remove pivot bolts to release the suspension links

3.3 Displace and clean suspension components ready for examination

3.5 Grease bushes during assembly. Do not omit dust seals

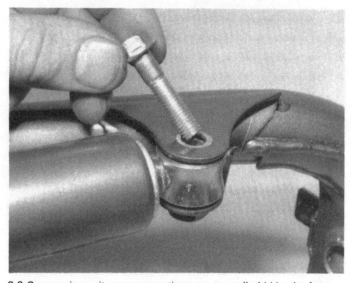

3.6 Suspension unit upper mountings are normally hidden by front mudguard (see text)

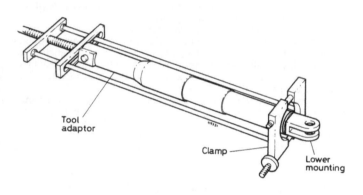

Fig. 4.2 Suspension unit spring compression tool (front suspension unit shown)

4 Rear suspension pivots: examination and renovation

1 Rear suspension is provided by arranging the engine/transmission unit so that it is free to pivot around its front mounting point, the rear of the assembly being controlled by a single coil spring and shock absorber unit. A mounting plate serves to link the engine crankcase to a point on the frame. This plate is of pressed-steel construction, its forward end containing two bonded-rubber bushes. The plate is attached to the crankcase by a bolt which passes through the rear of the plate and through bonded-rubber bushes which are pressed into lugs cast into the crankcase. To prevent metal-to-metal contact between the mounting plate and the crankcase, a rubber buffer is fitted between the two components.

2 Should wear have developed in the pivot assembly, it will be detected as side-to-side movement of the engine/transmission unit. This can be checked by grasping the rear of the transmission casing and pushing it from side to side whilst ensuring that the machine is held steady. Any discernible play will necessitate prompt attention before the resulting poor handling causes an accident.

3 Although it is possible to remove the mounting plate without the engine/transmission unit being removed completely from the machine, it is more satisfactory in practice to withdraw the unit completely so that proper attention can be paid to the bushes pressed into the crankcase lugs. Full instructions for unit removal are contained in Section 4 of Chapter 1.

4 Removal of the mounting plate from the crankcase, with the engine/transmission unit in or out of the frame, is a straightforward procedure. If the unit has remained in the frame, position a support between the crankcase to prevent the unit from dropping and then unscrew the locknut from the end of the plate to frame attachment bolt. With the locknut and plain washer removed, use a soft-metal drift and hammer to drift the bolt from position.

5 Remove the nut and washer from the end of the plate to crankcase attachment bolt and drift the bolt from position. The mounting plate can now be detached and placed to one side, ready for examination.

6 Carry out a close examination of the bonded-rubber bushes contained in both the mounting plate and the crankcase lugs. Each bush is made up of a rubber cylinder which is bonded to an external and an internal steel sleeve. Inspect each bush for deterioration of the rubber, wear of the inner sleeve and for separation between the rubber and either sleeve. If any such fault is evident, bush renewal is required.

7 The bushes fitted to the crankcase are a tight drive fit in their housing lugs. When the time comes for renewal, it will probably be found that, due to corrosion between the dissimilar metals (aluminium

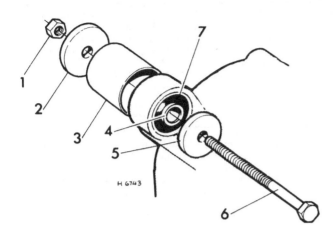

Fig. 4.3 Crankcase bush removal drawbolt tool

1	Nut	5	Washer
2	Washer	6	Drawbolt
3	Tube	7	Bush
4	Inner sleeve		

and steel), the already tight bushes have become almost immovable. As a means of removal, attempting to drive the bushes out will probably prove unsuccessful, because the rubber will effectively damp out the driving force, and damage to the lugs may occur. It is suggested that the bushes are drawn from position using a fabricated puller as shown in the accompanying diagram. This can be made from a short length of thick-walled tube, the inside diameter of which is slightly larger than the outside diameter of the bush, and two thick plate washers, one of which has an outer diameter slightly smaller than that of the bush outer sleeve and the other having a diameter which is greater than that of the tube. It is advisable to use a high tensile bolt and nut, which will be better able to take the stresses involved.

8 Heating the crankcase lugs in boiling water to expand the alloy and introducing penetrating oil around the bush to lug joint are both methods of helping to free the bushes before applying force to them by using the fabricated puller. If, having used every means possible, it is found that the bushes are reluctant to move, it is recommended that the engine/transmission unit be returned to a Honda dealer whose expertise can be brought to bear on the problem. Do not attempt any method of bush removal that may crack or damage the crankcase lugs.

9 Removal of the bushes fitted in the mounting plate can be achieved by supporting the plate on wooden blocks so that one of the bushes can be drifted downwards and out of its location. To do this, insert a long metal drift through the centre of the opposite bush so that its end abuts against the edge of the outer sleeve of the bush to be removed. Work the drift around the bush sleeve so that the bush remains square to the mounting plate during removal. A fairly heavy hammer will be needed to strike the drift in order to effect the initial freeing of the bush. Again, the initial use of heat and penetrating oil will aid removal of the bush. It must be noted that the bushes fitted in the mounting plate are not listed as separate items to the plate itself. Rather than incur the expense of buying a complete mounting plate assembly, it is well worth contacting a motor factor or any of the suppliers of bearings and bushes who advertise in the Yellow Pages or motoring trade magazines as to whether a similar type of bush can be supplied.

10 All new bushes may be driven into their locations by using a tubular drift against the outer sleeve. If this method is used, make sure that the opposite end of the bush location is well supported. In the case of the crankcase bushes, reverse the operation of the puller to draw each bush into position. Take care to check that both the bush outer surface and the bore of the bush location are free from all traces of corrosion and are cleaned of all contamination. Smearing a thin film of grease over the surface of the location bore will aid insertion of the bush. Take care to remove all excess grease from around the bush after insertion.

11 Continue the examination sequence by inspecting the mounting plate, the crankcase lugs and the frame mounting points for signs of fatigue or failure. Any fatigue in the crankcase lugs may well be indicated by a series of very fine hairline cracks. It is unlikely that such cracking will be seen unless the lugs have been thoroughly cleaned and degreased. Any ovality of the holes in the frame mounting points will necessitate specialist work by a skilled welder and machinist, who will have to fill the holes with metal before redrilling them to accept an unworn pivot bolt.

12 Inspect the shafts of both attachment bolts for signs of wear. This is not likely to be found on the crankcase to mounting plate bolt but may well be seen in the form of two grooves where the frame to mounting plate bolt forms a contact with the frame mounting points.

13 Align the unit mounting plate with the centre of each crankcase mounting and carefully drift the rear mounting bolt into position. Applying a light coating of grease to the shank of this bolt will ensure that it does not become corroded to the centre of the bonded rubber bushes, thus making extraction of the bolt at a later stage relatively trouble free. Fit the nut and washer to the rear mounting bolt and tighten the nut to a torque loading of 25 – 33 lbf ft (3.5 – 4.5 kgf m). Refer to Chapter 1 for full details of refitting the engine/transmission unit into the frame.

4.6 Engine mounting plate can be removed for examination

5 Rear suspension unit: examination and renovation

1 The rear suspension unit can be removed from the machine simply by unscrewing the bolt which retains it to the transmission casing and then unscrewing the single flange nut which retains it to the frame mounting. This operation must be carried out with the machine supported properly on its centre stand so that the transmission casing is allowed to drop away from the suspension unit once the retaining bolt is removed. Pull the lower mounting of the unit rearwards to clear the transmission casing and then pull the unit sideways clear of the frame.

2 The suspension unit comprises a hydraulic damper (effective primarily on rebound), a concentric spring and a rubber stop. The unit is secured to the frame and transmission casing through rubber-bushed lugs. The top lug of the unit contains a rubber insert which can be displaced fairly easily by the application of finger pressure. The lower bush is a press fit into the transmission casing and is more difficult to remove; various methods of bush removal are described in the previous Section.

3 Carry out a careful examination of the suspension unit whilst noting the following points. Inspect the damper unit for signs of leakage, corrosion or damage to the chromed surface of the piston shaft, deterioration of the seal, rubber buffer and rubber mounting

bushes, and damage to the piston housing. The damper unit can only be checked for efficiency after the suspension unit has been dismantled. There is no means of draining or topping up the fluid in the unit because it is sealed during manufacture. If signs of fluid leakage are apparent, the complete damper assembly must be renewed.

4   Check also for straightness of the damper rod. The damper unit must be renewed if this rod is seen to be bent. Breakage of the concentric spring will be obvious and renewal of the spring will necessitate its detachment from the damper unit.

5   The suspension unit can be dismantled by compressing the spring against the top of the damper unit and then removing the lower mounting attachment from the threaded end of the damper rod. Honda supply a special tool for doing this job. This tool should be used in conjunction with the two special adaptors also supplied by Honda. If none of these tools can be borrowed or hired from an official Honda dealer, or the unit cannot be returned to the dealer for renewal of the defective component, it is recommended the unit be renewed. In the absence of the spring compressor tools it is not practicable to attempt

dismantling without the risk of personal injury.

6   For those having access to suitable equipment, note that the spring service limit is 201.8 mm (7.95 in), and that during assembly the tighter coils must face uppermost. Use Loctite or similar on the locknut threads and tighten to 1.5 – 2.5 kgf m (11 – 18 lbf ft).

7   The unit is installed by reversing the removal sequence. Note that the upper mounting nut should be tightened to 3.0 – 4.5 kgf m (22 – 33 lbf ft) and the lower mounting bolt to 2.0 – 3.0 kgf m (14 – 22 lbf ft).

6   Frame: examination and renovation

1   If the machine is stripped for a complete overhaul, this affords a good opportunity to inspect the frame for cracks or other damage which may have occurred in service. Check the points at which the lower section of frame tube joins both the steering head and the pressed steel rear section of frame; these are the points where fractures

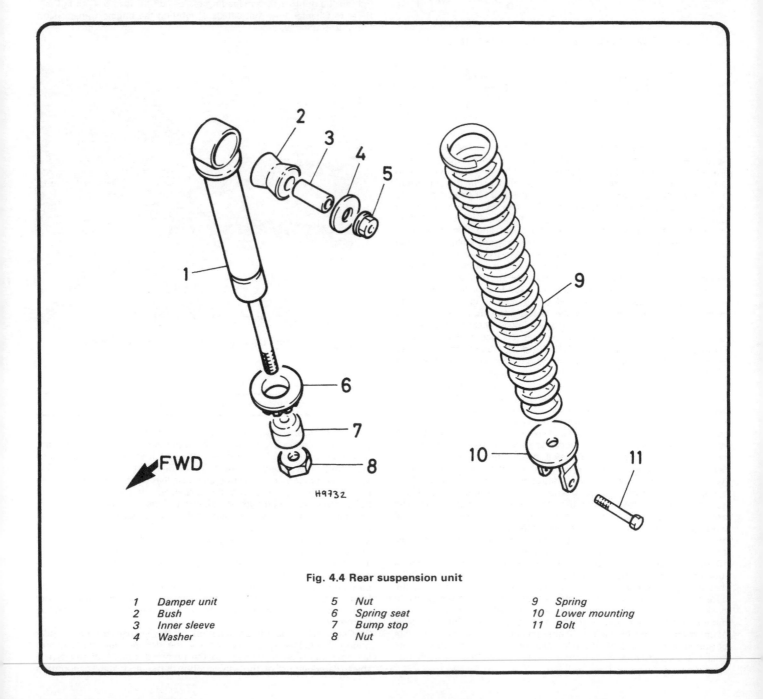

FWD

H9732

Fig. 4.4 Rear suspension unit

1	Damper unit	5	Nut	9	Spring
2	Bush	6	Spring seat	10	Lower mounting
3	Inner sleeve	7	Bump stop	11	Bolt
4	Washer	8	Nut		

are most likely to occur. Checking alignment of the steering head tube with the vertical section of the pressed steel section of frame will show whether the machine has been involved in a previous accident.

2 Check carefully areas where corrosion has occurred on the frame. Corrosion can cause a reduction in the material thickness and should be removed by use of a wire brush and derusting agents. After the machine has covered a considerable mileage, it is advisable to examine the frame closely for signs of cracking or splitting at the welded joints.

3 If the frame is broken or bent, professional attention is required. Repairs of this nature should be entrusted to a competent repair specialist, who will have available all the necessary jigs and mandrels to preserve correct alignment. Repair work of this nature can prove expensive and it is always worthwhile checking whether a good replacement frame of identical type can be obtained at a reasonable cost.

4 Remember that a frame which is in any way damaged or out of alignment will cause, at the very least, handling problems. Complete failure of a main frame component could well lead to a serious accident.

5.1 Rear suspension unit is easily unbolted for renewal

7 Body panels: removal and refitting

1 To gain access to most of the service items on the machine it is first necessary to remove one or more of the body panels. To avoid repetition, the procedure for removing each panel is described below. Where a particular operation requires that a panel be removed before work commences this is mentioned in the text, and this section should be consulted for details.

2 The various panels are of plastic construction and should be handled carefully. Whilst fairly strong, the surface finish is easily damaged if handled carelessly, and the locating tabs in particular are easily broken. Never force any panel; check that all of the fasteners have been removed and that removal is being tackled in the right sequence. Once detached, place the panel on some soft cloth well away from the working area. Keep all panels away from brake fluid, solvents and thinners, any of which will damage the surface.

Left-hand side panel

3 Unlock and open the seat. Remove the two bolts and two domed nuts which secure the rear carrier and lift it away, taking care not to scratch the bodywork. On the NB50 Vision-X model, take care not to lose the two spacers fitted to the carrier mounting studs. Release the single domed nut below the front of the seat and the two screws at the lower edge of the left-hand side panel. Carefully remove the side panel and place it to one side.

4 In the case of the NE50 Vision model, remove the carrier as detailed above, then remove the single domed nut below the front of the seat,

the screw immediately below it, and the single screw at the front lower edge of the side panel. Disengage the panel at the rear, taking care to avoid damaging the locating tabs, then swing it upwards and outwards until it can be freed at the front lower corner.

Right-hand side panel

5 Remove the left-hand side panel as detailed above. On the NB50 Vision-X model, open the luggage locker in the right-hand side panel and remove the single recessed bolt just below the latch plate. On both models, remove the screw(s) at the lower edge of the panel and lift it clear of the frame.

Front cover

6 The legshield assembly is a double-skinned structure comprising the front cover and the legshield mouldings. These are held together by ten screws fitted from the inner edge of the legshield. To release the front cover, remove the ten screws securing it to the legshield and then release the three domed nuts which secure the cover to the frame. The cover can then be lifted away exposing the legshield and much of the wiring, including the main multi-pin connector which forms the connection point for the majority of the electrical components and switches.

Floorboard

7 Remove the left-hand and right-hand side panels (see above). Unscrew the four bolts which retain the floorboard to the outrigger brackets beneath it and manoeuvre it clear of the frame.

Legshield

8 Remove the side panels, floorboard and front cover as detailed above. Release the turn signal relay from its bracket on the legshield and remove the single screw to free the multi-pin connector. On later models disconnect the front turn signal wiring.

9 Moving to the inside of the legshield, release the two screws which secure its lower edge to the frame outriggers. Unlock the lid of the front luggage compartment and remove the single retaining bolt. The legshield can now be lifted away and placed to one side.

Handlebar nacelle

10 The handlebar assembly, instrument panel and switches are housed in a moulded plastic nacelle formed by two sections. It is rarely necessary to remove both of these, removal of the rear section normally proving adequate for access purposes. Start by releasing the three screws which secure the rear section of the nacelle. Lift it upwards, easing the speedometer cable through the steering column until it can be detached at the upper end. If necessary, the cable should first be released at the wheel to obtain sufficient free play.

11 The rear section of the nacelle can be lifted away after the instrument panel and switch wiring has been disconnected. In the case of the latter, this should be done at the main multi-pin connector, access to which will require the removal of the front panel (see above). It is advisable to number each of the connector strips and also the corresponding position on the connector block so that it can be refitted in its correct position. Once the wiring has been freed, lift the nacelle rear section away and place it to one side.

12 It is normal practice to continue by dismantling the remaining handlebar controls, the turn signals (early models), the headlamp so that the front section of the nacelle can be removed. In almost all cases this work can be avoided if care is taken. Where removal of the front section is unavoidable, however, remove the rear section and the front cover as detailed above, and also the rear view mirrors. Remove the single screw beneath the headlamp and the two mounting screws to the rear of the headlamp. Remove the turn signal lamp lenses, then release the lamps themselves. The front section can now be lifted away, together with the headlamp unit.

Reassembly

13 In most respects, the various body panels are refitted by reversing their removal sequence. Note that this extends to the order of removal; eg, the side panels and floorboard should not be fitted until the legshield is in place. Make sure that no undue force is used during installation and that any spacers and washers are refitted correctly to avoid straining the panels. Be particularly careful to ensure that all locating tabs are positioned correctly so that they hold the panel edges in alignment. It will be obvious from examination if they are misaligned.

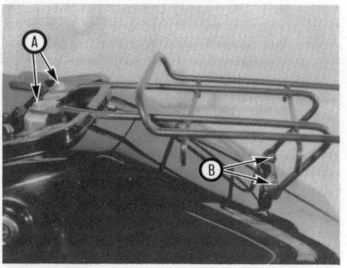

7.3a Rack mounting bolts (A) and domed nuts (B)

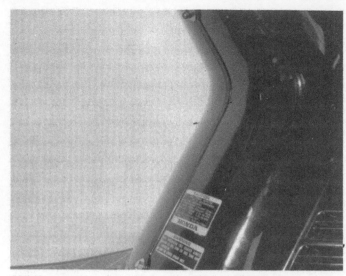

7.3b Remove domed nut (top) and single bolt (bottom)

7.4a Note screw securing lower edge of side panel

7.4b Take care not to damage locating tabs at front of panel

7.4c More locating tabs are located on rear seam between panels

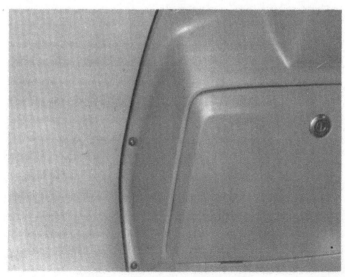

7.6a Front panel is retained by ten cross-head screws ...

7.6b ... and by three domed nuts

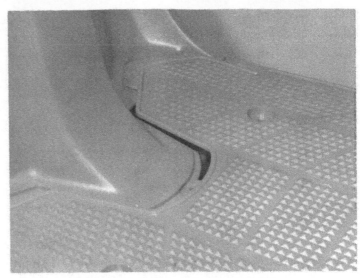

7.7a Remove footboard bolts and pull footboard back ...

7.7b ... noting locating tabs near front edge of panel

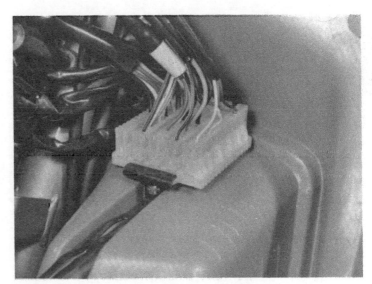

7.8 Multi-pin connector is secured by bracket and single screw

7.9a To release door, disengage the side stays ...

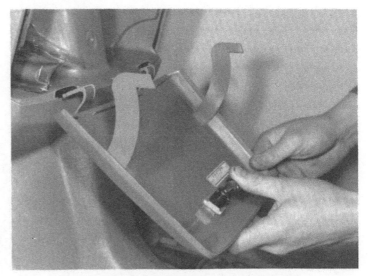

7.9b ... and lower door until hinges come free of legshield

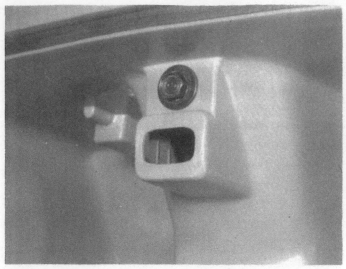

7.9c Single bolt is located inside glove box recess

7.9d Lower mounting screws are normally hidden under footboard

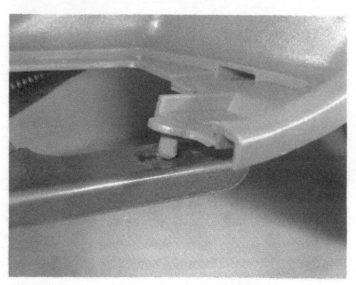

7.9e Note locating pegs at lower edge of legshield

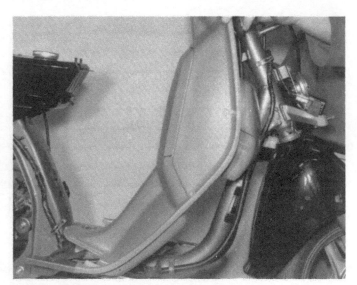

7.9f Legshield can be lifted up and back as shown

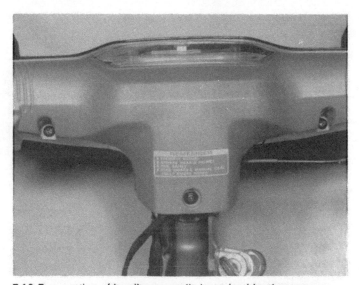

7.10 Rear section of headlamp nacelle is retained by three screws

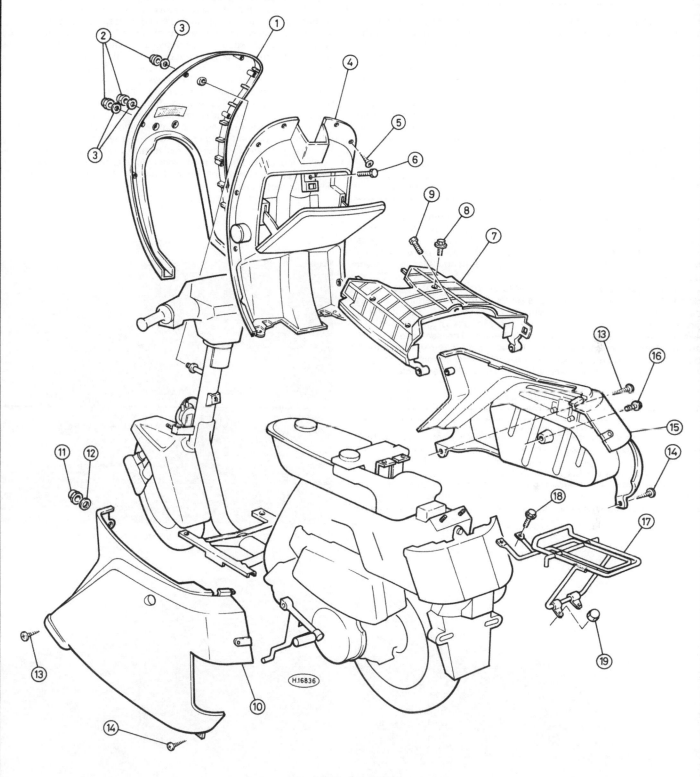

Fig. 4.5 Body panels

1	Front cover	6	Bolt	11	Domed nut	16 Bolt
2	Domed nut – 3 off	7	Footboard	12	Washer	17 Rear carrier
3	Washer – 3 off	8	Bolt – 4 off	13	Screw – 2 off	18 Bolt – 2 off
4	Legshield	9	Bolt	14	Screw – 2 off (NB only)	19 Domed nut – 2 off
5	Screw – 10 off	10	Left-hand side panel	15	Right-hand side panel	

Note: NB50 Vision-X side panels shown

8 Seat: removal and refitting

1   The seat is mounted above the fuel tank. It is retained in position by a hinge which is secured to the front section of the seatpan and to the front of the fuel tank. Two rubber buffers prevent the seatpan from coming into direct contact with the rear of the fuel tank when the seat is lowered and a seat lock is provided to secure the seat in its down position. Note that the seat is designed to take the weight of no more than one person.

2   To remove the seat, unlock and raise it. Unscrew the two flange nuts which retain the seat hinge to the fuel tank and then lift the seat clear of the machine. Fitting the seat is a direct reversal of removal procedure. Check that the seat is correctly aligned on the fuel tank before finally tightening the two securing nuts.

9 Steering lock and ignition switch: location and renewal

1   The steering lock and ignition switch are combined in one unit which is mounted on a plate which forms part of the steering head assembly. When the key is turned to its 'Lock' position, a bar projects from the lock body and through a cutout in the lower periphery of the handlebar centre bracket, thus locking the handlebars in one set position.

2   To gain access to the lock it is necessary to remove the legshield as detailed in Section 7. Once this is removed the lock retaining screws can be reached, as can the wiring connector. Note that it is not practicable to repair the lock or switch sections of the assembly; if either proves defective, it must be renewed. The seat and luggage compartment locks are opened by the same key, so consideration must be given to renewing all locks as a set. The latter simply fit through a hole in the panel and are secured by a spring clip (see photograph).

10 Stand: examination and maintenance

1   The Honda Vision models are equipped with a centre stand which is formed out of tubular steel. The stand pivots on a steel shaft which passes through two plates on the frame structure and is retained in position by a plain washer and a split pin.

2   Maintenance of the stand consists of checking that it retracts smoothly and fully against its stop. The stand retracts by a spring connected between the stand and the frame structure. Closely examine this spring for any signs of fatigue or wear at the points where it passes through the stand or the frame. If the stand is slow to return against its stop, it is likely that its pivot points are in need of lubrication and probably need cleaning of any hardened deposits or corrosion.

3   Should the stand require any attention, it can be removed by removing the split shaft and then displacing the pivot shaft. Once this is done, the spring can be unhooked from the stand and the stand manoeuvred clear of the machine. Check for wear at the bearing surfaces of the shaft to stand. Excessive wear at these points will affect the efficiency of the stand in retracting. The same applies if excessive wear is seen at the spring retaining holes.

4   Note that fitting of the stand can prove rather arduous because it is necessary to manoeuvre the stand into position against spring pressure whilst the pivot shaft is fitted. There is no easy way of accomplishing this operation, and a mixture of patience and strength is essential. It is almost indispensable to have some assistance during reassembly. Remember to fit a new split pin and to retain it in position by correctly bending apart its legs.

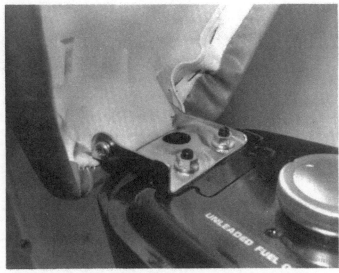

8.2 Seat hinge is secured by two flange nuts to top of fuel tank

9.2a Panel locks are a push fit in their mounting holes ...

9.2b ... and are held in position by spring clips

11 Kickstart lever: examination and renovation

1 The kickstart lever is secured to its splined shaft by means of a pinch bolt which passes through the edge of its mounting boss. The inner surface of the lever boss is splined to correspond with the shaft. The footrest section of the lever is designed to swivel about the end of the main section of lever, so that when the engine is started, it can be tucked out of harm's way against the transmission casing.

2 A spring-loaded ball bearing locates the footrest section of the lever in either the operating or folded position. If the action of this mechanism becomes sloppy it is probable that the spring behind the ball bearing needs renewing. It is therefore necessary either to purchase a new item or enquire at a motorcycle breaker as to the availability of a second-hand serviceable item.

3 If the action of the folding mechanism is stiff, apply a penetrating fluid to free it before lubricating it with clean motor oil. Do not ride the machine with the footrest section of the lever in the operating position as it is possible that in the unfortunate occurrence of the machine being 'dropped', the end of the footrest section striking the ground will serve to push the main section of lever into the transmission casing with the resulting breakage and expense.

4 If the kickstart lever is bent in an accident, it should be removed and straightened by using the following procedure. Unscrew and remove the pinch bolt from the lever boss and pull the lever off its shaft. Clamp the lever in between the jaws of a vice and heat the area around the deformed section to a dull red heat by using the flame of a blowlamp. Straighten the lever and allow it to cool naturally. Cooling the lever by immersing it in cold water is not only a dangerous practice but will cause the affected section of shaft to become embrittled.

5 If there is evidence of failure in the metal of the lever either before or after straightening, it is essential that the damaged component be renewed. If the kickstart lever breaks whilst being used, serious injury could result to the leg or foot. When fitting the lever, check that the alignment dot on the end of its shaft is adjacent to the gap in the lever boss and ensure that the lever is not in contact with the transmission casing at any point along its length.

12 Speedometer head: removal and refitting

1 The instrument head itself is generally reliable, and is the least likely culprit in the event of failure; this normally being attributable to the cable rather than the instrument mechanism. If, however, it is noted that the speedometer has ceased to function whilst the odometer (mileage recorder) still functions, the instrument can be assumed to have failed. No form of repair is practicable at home, and a replacement speedometer will be required. The only alternative is to seek the assistance of one of the companies who specialise in this type of repair work.

2 The speedometer head is removed together with the rear section of the handlebar nacelle as described in Section 7 of this Chapter. Once this is removed, the screws retaining the instrument can be released to permit its removal. Further dismantling of the unit will allow the fuel gauge to be separated from the speedometer; the two instruments can be obtained individually.

11.4 Kickstart lever is retained by a pinch bolt

13 Speedometer drive and cable: examination and maintenance

1 Drive from the speedometer gearbox is transmitted to the instrument head by way of a flexible cable. This flexible cable consists of an inner cable which runs inside, and is protected by, a reinforced outer cable. This arrangement allows the drive to pass through gentle bends and absorbs the relative movement between the front wheel and the instrument.

2 Although considered to be flexible, it is preferable to ensure that the cable does not pass through acute bends, which would shorten its effective life. The straighter the cable's run, the lower the rate of wear and risk of breakage.

3 If the speedometer ceases to function, suspect a broken cable. Inspection will show whether the inner cable has broken.

4 Spin the inner cable to check for resistance. Most cables have a tight spot, but if the resistance is severe and a wavering speedometer has been noted, the cable should be renewed. Lubrication is difficult with this type of cable, but an aerosol chain grease or a silicone-based lubricant can often be introduced using the aerosol's thin extension nozzle. Do not apply excess lubricant to the upper end of the cable otherwise there is a risk of it working up into the instrument head.

5 Drive to the cable takes the form of a gear mechanism built into the front brake backplate. This is not normally prone to wear, and maintenance can be restricted to inspection and regreasing whenever the front brake components receive attention. Any mechanical failure will be obvious and will require renewal of the damaged parts.

Chapter 5 Wheels, brakes and tyres

Refer to Chapter 7 for information relating to the SA50 Met-in

Contents

General description 1
Front wheel: examination and renovation 2
Front wheel: removal and refitting 3
Front wheel bearings: removal, examination and refitting 4
Speedometer drive: removal, examination and refitting 5
Rear wheel: examination and renovation 6

Rear wheel: removal and refitting 7
Rear wheel bearings: general 8
Brakes: adjustment and checking for wear 9
Brakes: examination and renovation 10
Tyres: removal and refitting 11
Valve cores and caps 12

Specifications

Wheels

Type	Pressed steel
Rim runout (max) – axial and radial	2.0 mm (0.08 in)
Front wheel spindle runout (max)	0.2 mm (0.01 in)

Brakes

Type	Single leading shoe (sls) drum
Brake lining thickness:	
Front	3.5 mm (0.14 in)
Service limit	1.5 mm (0.06 in)
Rear	4.0 mm (0.16 in)
Service limit	2.0 mm (0.08 in)
Brake drum diameter:	
Front	80.0 mm (3.15 in)
Service limit	80.5 mm (3.17 in)
Rear	95.0 mm (3.74 in)
Service limit	95.5 mm (3.76 in)

Tyres

Size – front and rear	2.75-10-4PR
Pressures:	
Front	21 psi (1.5 kg/cm²)
Rear	28 psi (2.0 kg/cm²)

Torque wrench settings

Component	kgf m	lbf ft
Front wheel spindle nut	4.0 – 5.0	29 – 36
Rear wheel stub axle nut..	8.0 – 10.0	58 – 72
Brake arm bolt – front and rear	0.4 – 0.7	3 – 5

1 General description

The Honda Vision employs small diameter painted pressed steel wheels with an integral brake drum and wheel centre welded to the rim. The front wheel is of normal motorcycle fitting with a wheel spindle passing through the fork ends, whilst the rear wheel has a splined boss which engages a corresponding spline on the output shaft or stub axle of the reduction gearbox.

A simple single leading shoe (sls) drum brake is fitted to both wheels, the brakes being controlled by handlebar levers in a similar arrangement to that found on bicycles. Although similar in size and construction the wheels are not interchangeable.

Tyre sizes are 2.75-10-4PR front and rear and are of conventional tubed type.

2 Front wheel: examination and renovation

1 Place the machine on its centre stand and place a large wooden block, or similar, beneath the footboard panel mounting points so that the front wheel is raised clear of the ground.

2 Spin the wheel and check for rim alignment in both the axial and radial planes by placing a pointer close to the rim edge. If the total alignment variation is greater than the given service limit of 2.0 mm (0.08 in), it is the manufacturer's recommendation that the wheel be renewed. This is, however, a counsel of perfection and in practice a larger amount of runout may not affect the handling qualities of the machine to an excessive degree.

3 When cleaning the machine, examine the wheel carefully. Any areas where the paint finish has been chipped away must be refinished at the earliest available opportunity if the likelihood of the wheel rusting is to be avoided. Any rust already present on the wheel should be removed by rubbing with emery cloth until the exposed metal is bright and clean and then coating the area with one of the rust removing agents sold by any of the well known motor accessory stockists before applying the final paint finish.

4 Inspect the complete wheel for localised damage in the form of cracks or dents. It is possible that a dent can be ignored as long as the paint finish has remained undamaged, but a crack will render the wheel unfit for further use unless it is found that a permanent repair is possible through welding. This method of repair requires a great degree of skill and therefore the advice of a wheel repair specialist should be sought.

5 Check closely that the area of wheel rim which supports the tyre

bead is in no way damaged. A tyre that is improperly seated and which moves around the wheel rim can only constitute potential danger.

6 Finally, check the wheel for any signs of play in the wheel bearings. This may be indicated by being able to induce side-to-side movement of the wheel whilst the front forks are held securely in one position. Any roughness heard or felt in the bearings as the wheel is rotated will mean that they are in need of immediate renewal.

3 Front wheel: removal and refitting

1 Place the machine on its centre stand and raise the front wheel clear of the ground by positioning a large wooden block, or similar, beneath the footboard panel mounting points.

2 Removal of the wheel is a simple and straightforward procedure which should begin with the detachment of the brake operating cable from the brake cam lever. To do this, unscrew fully the adjuster nut, and disengage the cable from the brake arm by pulling it through the trunnion. Refit the trunnion and nut on the cable end for safe-keeping.

3 When disconnecting the brake cable, note the condition of the spring and rubber gaiter fitted over the cable end. The gaiter acts to protect the cable inner from contamination by road salts and dirt. Renew the gaiter if it is split or perished. If the spring is seen to be broken or weakened by corrosion, then it also should be renewed.

4 Disconnect the speedometer drive cable from its location in the brake backplate by first removing the small cross head securing screw and then pulling the cable clear. Allow the cable to hang clear of the wheel.

5 Unscrew the flanged locknut from the end of the wheel spindle and carefully drift the spindle from position with the wheel supported. Use a soft-metal drift in conjunction with a hammer and take great care not to damage the threaded end of the spindle. The wheel can now be manoeuvred clear of the front forks.

6 Before fitting the wheel, clean the spindle of any hardened grease or corrosion and check it for straightness. Honda recommend that the spindle is supported between two V-blocks and its runout checked by means of a dial gauge positioned at the spindle centre. The runout measured should not exceed the service limit of 0.2 mm (0.01 in). It will suffice, however, to hold the spindle against a good straight-edge to check for straightness. A bent spindle must be replaced with a serviceable item. After having cleaned and checked the spindle, smear a light coating of a high melting-point, lithium based grease along its length.

7 Fitting the front wheel is a direct reversal of the removal procedure, whilst noting the following points. When lifting the wheel into position between the forks, ensure that the projection of the right-hand pivot arm fits cleanly into the slot cast into the brake backplate. Note that if the brake backplate is allowed to rotate, due to its not being held in position by the projection of the pivot arm, the wheel will lock on the first application of the front brake, with disastrous consequences. The collar must be located in the oil seal.

8 Fit the wheel spindle with its retaining nut and tighten the nut to the specified torque loading of 29 – 36 lbf ft (4.0 – 5.0 kgf m). After having relocated the speedometer drive cable, secure it in position with the screw and then spin the front wheel whilst observing the instrument needle, to check that drive is being transmitted from the wheel to the instrument head.

9 Reconnect the brake cable and wind the adjuster nut forward until the amount of free play measured at the top of the handlebar lever is 10 – 20 mm (3/8 – 3/4 in). Check for correct operation of the brake by spinning the wheel and applying the brake lever. If, when the brake is off, the brake shoes are heard to be brushing against the surface of the wheel drum, back off the adjuster nut slightly until all indication of binding disappears. If the brake shoes have been renewed, the brake will need to be readjusted after a period of bedding-in has been allowed.

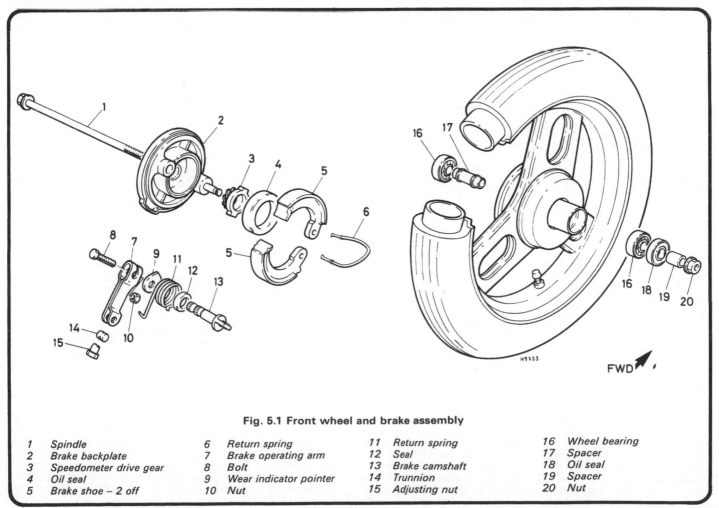

Fig. 5.1 Front wheel and brake assembly

1	Spindle	6	Return spring	11	Return spring	16	Wheel bearing
2	Brake backplate	7	Brake operating arm	12	Seal	17	Spacer
3	Speedometer drive gear	8	Bolt	13	Brake camshaft	18	Oil seal
4	Oil seal	9	Wear indicator pointer	14	Trunnion	19	Spacer
5	Brake shoe – 2 off	10	Nut	15	Adjusting nut	20	Nut

3.2 Unscrew the adjuster nut to free cable from brake arm

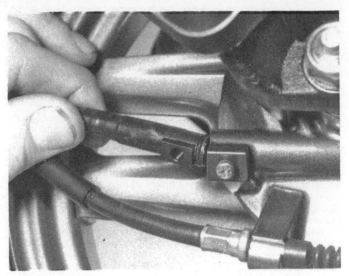

3.4 Speedometer cable is secured by a single cross-head screw

3.5 Remove the wheel spindle and lift wheel clear of forks

4 Front wheel bearings: removal, examination and refitting

1   There are two bearings in the hub of the front wheel. If it is possible to induce any side play in the wheel with it fitted to the machine, or to feel any roughness in the bearings as the wheel is rotated, both wheel bearings must be renewed.

2   Before bearing removal can be undertaken, the brake backplate assembly must be removed from the wheel hub and the wheel laid flat on a work surface, supported by wooden blocks so that enough clearance is left beneath the wheel to drive the bearing out. To lessen the risk of distortion to the wheel, ensure that these blocks are placed as close to the bearing housing as possible.

3   With the brake backplate side of the wheel hub facing uppermost, place the end of a small flat-ended drift against the upper face of the lower bearing and tap the bearing downwards out of the wheel hub. The spacer located between the two bearings may be moved sideways slightly in order to allow the drift to be positioned against the face of the bearing. Move the drift around the face of the bearing whilst drifting it out of position, so that the bearing leaves hub squarely.

The oil seal will be displaced along with the bearing. Note the spacer collar located in the centre of this seal.

4   With the one bearing removed, the wheel may be lifted and the spacer withdrawn from the hub. Invert the wheel and remove the second bearing, using a similar procedure to that used for the first.

5   Wash both bearings, the bearing spacer and their hub location thoroughly in clean petrol to remove all traces of the old grease. Remember to take the necessary fire precautions whilst doing this. Check the bearing tracks and balls for wear, pitting or damage to the hardened surfaces. A small amount of side movement in the bearing is normal but no radial movement should be detectable. If wear or damage is found the bearing in question should be renewed.

6   If the original bearings are to be refitted, then they should be re-packed with the recommended grease before being fitted into the hub. New bearings must also be packed with the recommended grease. Ensure that the bearing recesses in the hub are clean and both bearing and recess mating surfaces lightly greased. The two bearings and central spacer may now be fitted. With the hub well supported by the wooden blocks, drift the bearing into the brake backplate side of the wheel hub using a length of metal tube, or a socket of suitable diameter and a soft-faced mallet. Invert the wheel, insert the spacer and fit the second bearing using the same procedure as that given for the first. Note that both bearings must be fitted with their sealed sides facing outwards.

7   It is considered good practice to renew the oil seal every time the bearings are removed. In any event, the seal must be renewed if it is seen to be damaged or if it has begun to deteriorate. With the seal pressed partially into its recess in the wheel hub, tap it home by using a socket or length of metal tube of suitable diameter in conjunction with a soft-faced hammer. Lightly grease the spacer collar and insert it into position through the seal.

5 Speedometer drive: removal, examination and refitting

1   The speedometer drive assembly is contained within the front wheel brake backplate and should be examined and repacked with grease whenever work is carried out on the wheel bearings or brake assembly.

2   To remove the main drive gear from its housing, place the brake backplate assembly, inner side uppermost, on a work surface. Remove the brake shoes and spring, as described in Section 10, and inspect the large dust seal for signs of damage and deterioration. To renew this seal, carefully lever it out of position using the flat of a screwdriver; great care must be taken not to damage the surrounding alloy casting. A new seal may be fitted after removal, examination and fitting of the drive gear assembly.

3 The main drive gear can now be withdrawn from its housing. In practice, some difficulty may be experienced in withdrawing the gear due to it being engaged with the worm drive gear and to the retentive qualities of the grease around the base of the gear.

4 Remove all old grease from the main drive gear and from its housing by wiping the components with a clean rag. It is unlikely that the main drive gear will wear to any excessive degree until the machine has covered a considerable number of miles. Nevertheless, inspect the gear for broken teeth, broken drive tabs or any sign of excessive wear. If the gear is seen to be defective, it must be renewed. Carry out a similar inspection on the worm drive gear. Unfortunately, Honda do not list this component as a separate item to the brake backplate so it must be assumed that any defective worm drive gear will necessitate replacement of the complete backplate. Return the component to an official Honda dealer for further advice on this matter.

5 To reassemble the speedometer drive gear, pack the housing with a high melting-point, lithium-based grease before inserting the main drive gear into position. Where required, fit the new dust seal into its recess in the brake backplate by pressing it into position evenly and squarely. Wipe a small amount of grease over the lip of this seal. The brake shoes and spring may now be refitted and the brake backplate assembly inserted into the wheel hub. Take care to ensure that the drive tabs of the main drive gear are aligned correctly with the corresponding slots in the wheel hub boss.

4.6a Fit the bearing and oil seal, then turn wheel over ...

4.6b ... and fit the bearing spacer

4.6c Place new bearing in position ...

4.6d ... and drive home squarely using a large socket

5.3 Remove the brake shoes and oil seal to gain access to the speedometer drive gear

5.5 Grease gear housing liberally, then refit oil seal ...

6 Rear wheel: examination and renovation

1   Place the machine on its centre stand and position a support beneath the engine crankcase to keep the rear wheel clear of the ground. With an assistant steadying the machine, check for wear in the wheel centre spline by gripping each side of the wheel and attempting to move it from side-to-side. Any movement between the wheel and its stub-axle will necessitate removal of the wheel for further investigation.
2   The remaining examination and renovation procedure is as for the front wheel (see Section 2, paragraphs 2 to 5). Rotate the wheel to check for roughness in the stub axle bearings.

7 Rear wheel: removal and refitting

1   Commence removal of the rear wheel by placing the machine securely on its centre stand and remove the exhaust system by following the instructions given in Chapter 2.
2   It is now necessary to prevent the rear wheel from rotating whilst the wheel securing nut is slackened. This is most easily accomplished by taking the machine off its stand and having an assistant sit astride it whilst at the same time applying both brakes to prevent the machine from moving forward. If an assistant is not available, it may be possible to lock the wheel sufficiently by turning the brake cable adjuster fully in so that the brake is applied. The wheel securing nut is tightened to a high torque loading and will therefore be difficult to loosen.
3   Remove the wheel securing nut. Position a support beneath the engine crankcase so that the back of the machine is prevented from dropping, grip each side of the wheel and pull it clear of its stub-axle and away from the machine.
4   With the wheel removed, take the opportunity to clean and inspect the splines of the stub-axle and of the wheel itself. Inspect the load bearing surfaces of each spline for signs of deterioration in its surface finish. This, and any wear lip found along the length of each spline, will necessitate renewal of both the shaft and the wheel. Carry out a final check for wear by temporarily refitting the wheel over the shaft and feeling for any movement between the two components. With both shaft and wheel considered to be serviceable, apply a very light smear of a high melting-point, lithium-based grease to the shaft splines before finally sliding the wheel over the shaft. Bear in mind that just enough grease is needed to prevent dryness and the resulting ingress of moisture between the spline surfaces. Over lubrication will lead to contamination of the brake shoes.
5   It is good practice to renew the wheel securing nut each time it is removed. This is because every locknut of this type loses some of its gripping qualities each time it is moved along the threaded section of shaft to which it is attached and at some stage it may unwind whilst the machine is in motion, with disastrous consequences. The price of a single locknut is nothing in comparison to that of a new transmission casing. If it is found impossible to obtain a nut, coat the threads of the shaft with a thread locking compound before fitting the nut with its washer and tightening it to the recommended torque loading of 58 — 72 lbf ft (8.0 - 10.0 kgf m). A plain washer is fitted under the nut on SA50-P models; ensure that this is refitted before installing the nut.
6   Refit the exhaust system by following the instructions given in Chapter 2. Before riding the machine, carry out a quick check to ensure that the rear brake is correctly adjusted. There should be 10 – 20 mm (^{3}/8 – ^{3}/4 in) of free play measured at the tip of the handlebar lever. The amount of free play can be adjusted by rotation of the adjuster nut at the wheel end of the cable. If the brake shoes are heard to be brushing against the surface of the wheel drum as the wheel is rotated, back off the adjuster nut slightly until all indication of binding disappears. If the brake shoes have been renewed, the brake may be readjusted after a period of bedding-in has been allowed. Check that the brake works efficiently and check the starter interlock system operation before taking the machine on the road.

8 Rear wheel bearings: general

Unlike the front wheel, the rear wheel is not fitted with its own bearings. The hub is splined and fits directly on the rear stub-axle, this being carried on bearings in the final drive casing. If play is detected, it will be necessary to examine these bearings for wear or damage, and reference should be made to Chapter 1 for details.

9 Brakes: adjustment and checking for wear

1   It is essential that the brakes on any motorcycle are always kept in correct adjustment. The procedure for carrying out adjustment of the brakes is both simple and straightforward and identical for each brake. Commence by checking the amount of free play measured at the handlebar lever end. If the amount of play measured is outside the set limit of 10 – 20 mm (^{3}/8 – ^{3}/4 in), it must be altered by turning the adjuster nut at the brake drum end of the cable. Check for correct operation of the brake by spinning the wheel and applying the brake lever. If, when the brake is off, the brake shoes are heard to be brushing against the surface of the wheel drum, back off the adjuster nut slightly until all indication of binding disappears. If the brake shoes have been renewed, the brakes should be readjusted after a period of bedding-in has been allowed.
2   With the brakes correctly adjusted, apply each handlebar lever fully and check the position of the indicator plate mounted on the brake camshaft with the arrow cast into the brake backplate. If the arrow on the indicator plate is seen to align with, or go past the arrow on the brake backplate, the brake shoe linings have worn beyond their service limit and should be renewed immediately.
3   After making adjustment of the rear brake check the starter interlock system. With the rear brake locked on check that the stop lamp bulb is illuminated and that the starter operates; with the rear brake off, the starter should not operate.

10 Brakes: examination and renovation

1   Both wheels incorporate a single leading shoe (sls) drum brake, each of which is operated by a handlebar-mounted lever.
2   To gain access to the components of each brake assembly, it is first necessary to remove the appropriate wheel as described in Sections 3 or 7. In the case of the front wheel, the brake backplate assembly is removed complete with the wheel from the machine; it can then be pulled clear of the wheel hub. The components of the rear brake assembly will remain attached to the transmission casing of the engine/transmission unit.
3   The brake drums should be checked for wear or damage after any accumulation of dust has been removed. This dust contains asbestos which can be harmful if inhaled. For this reason, **never** use

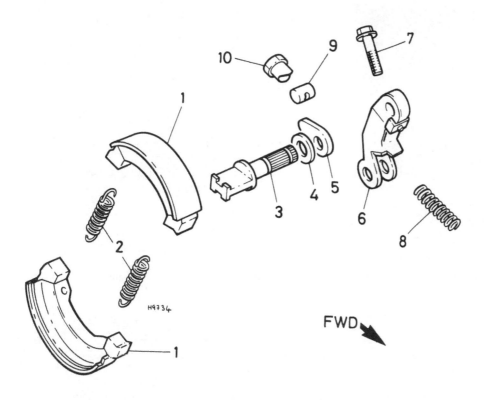

Fig. 5.2 Rear brake assembly

1. Brake shoe – 2 off
2. Return spring – 2 off
3. Camshaft
4. Seal
5. Wear indicator pointer
6. Operating arm
7. Bolt
8. Spring
9. Trunnion
10. Adjusting nut

FWD

compressed air to remove the dust. A petrol-moistened rag will remove the dust and clean the drum surface quickly and safely. Look for signs of scoring or any other damage on the drum surface. If badly damaged it may be possible to have the drum skimmed in a lathe by a suitably-equipped specialist. Failing this, renewal of the entire wheel is inevitable. In practice, such damage is unlikely to be encountered, and very light scoring can be accepted. Any rusting of the drum surface may be removed by careful use of fine abrasive paper.

4 The shoes can be checked for wear by measuring the thickness of the lining material at the point where the greatest amount of wear has taken place. The service limit for the thickness of the lining material is 1.5 mm (0.06 in) front brake, and 2.0 mm (0.08 in) rear brake.

5 The front brake shoes can be removed once the return spring has been released. This takes the form of a large C-shaped spring ring, the ends of which engage in holes in the shoe ends. To release the shoes, grasp the ends furthest from the brake cam, pulling them apart and away from the backplate. The shoes can be allowed to fold inwards to help in freeing the brake cam ends. It is recommended that eye protection is worn during this operation because the spring can jump free with unexpected force in an unpredictable direction.

6 The rear brake shoes have conventional extension coil springs fitted between the shoes. They can be freed by folding the shoes inwards until the ends can be disengaged from the cam and pivot. Once the shoes have been removed note the position of each spring in relation to the brake shoes before separating them.

7 If the existing shoes are to be refitted, they should be wiped with a clean petrol-moistened rag to remove the accumulated brake dust. The rear shoes in particular should be examined for signs of oil contamination. If this is slight it may be possible to remove the oil with petrol. An aerosol solvent known as Ultraclean, used mostly for electrical cleaning and degreasing, is ideal for this job and can be obtained from motor factors or through electrical shops. Severe contamination will necessitate renewal of the shoes. It follows that where oil contamination has been noted it will be necessary to rectify the leak before the brake is refitted. This is most likely to be a damaged oil seal on the stub axle and reference should be made to Chapter 1 for details.

8 Existing shoes that are to be refitted should also have their lining surface roughened sufficiently to break the glaze which will have formed in use. Glasspaper or emery cloth is ideal for this purpose but

take great care not to inhale any of the asbestos dust that may come from the lining surface.

9 Before fitting the brake shoes, check that the brake operating cam is working smoothly and not binding in its pivot. The cam may be removed by unscrewing the operating arm retaining bolt and pulling the arm off the shaft. The camshaft can then be extracted from its location in the brake backplate. Thoroughly clean the shaft and the bore in the brake backplate through which it passes. Clean and grease the shaft prior to reassembly and also lightly grease the faces of the operating cam. A light smear of high melting-point grease should also be placed on the face of the spigot about which the shoes pivot. The brake shoe wear indicator plate is aligned with the camshaft by matching the wide groove in the shaft end with the wide tooth of the plate. When fitting the cam operating arm, align the punch mark on the arm with that on the shaft. Note that the camshaft of the front brake incorporates a dust seal. If this seal is seen to be defective it must be renewed.

10 Finally, inspect the brake shoe springs for signs of damage or deterioration. After a considerable amount of use, the springs will take a 'set' and become inefficient. If it is suspected that this has happened, take the spring in question to an official Honda dealer and compare it with a new item. Note that the two springs of the rear brake must be of identical lengths and must always be renewed as a pair.

11 Check the spring to brake shoe contact points. If either component is excessively worn, it must be renewed. If the ingress of moisture into the brake drum has caused the springs to become corroded and heavily pitted, they should be renewed before subsequent failure occurs. Do not neglect to examine the rear brake cam operating arm return spring by using a procedure similar to that described above.

12 Reassemble and refit the brake assemblies by reversing the dismantling and removal procedures. With the rear brake assembly, holding the shoes in the V-shape ready to fit them onto the backplate and at the same time retaining the return springs, requires a little practice and a lot of patience. Once the ends of the shoes are located with the cam and pivot points however, it is easy to snap them into position by pressing downward. On no account use excessive force, otherwise there is a risk that the shoes may be permanently distorted, or the springs over-stretched. Do not forget to reset the brake adjustment or to test the brakes and starter interlock system for correct operation before taking the machine on the road.

10.2 Front brake backplate can be lifted away after wheel has been removed

10.5a Prise spring off groove on pivot pin and fold upwards

10.5b Shoes can now be worked off pivot and removed

10.6 Rear shoes mount on transmission casing and have conventional coil-type return springs

11 Tyres: removal and refitting

1   To remove the tyre from either wheel, first detach the wheel from the machine by following the procedure in Sections 3 or 7 of this Chapter, depending on whether the front or the rear wheel is involved. Deflate the tyre by removing the tyre insert and when it is fully deflated, push the bead of the tyre away from the wheel rim on both sides so that the bead enters the centre well of the rim. Remove the locking cap and push the tyre valve into the tyre itself.

2   Insert a tyre lever close to the valve and lever the edge of the tyre over the outside of the wheel rim. Very little force should be necessary; if resistance is encountered it is probably due to the fact that the tyre beads have not entered the well of the wheel rim all the way round the tyre.

3   Once the tyre has been edged over the wheel rim it is easy to work around the wheel rim so that the tyre is completely free on one side. At this stage, the inner tube can be removed.

4   Working from the other side of the wheel, ease the other edge of the tyre over the outside of the wheel rim that is furthest away. Continue to work around the rim until the tyre is free completely from the rim.

5   If a puncture has necessitated the removal of the tyre, reinflate the inner tube and immerse it in a bowl of water to trace the source of the leak. Mark its position and deflate the tube. Dry the tube and clean the area around the puncture with a petrol-soaked rag. When the surface has dried, apply the rubber solution and allow this to dry before removing the backing from the patch and applying the patch to the surface.

6   It is best to use a patch of the self-vulcanising type, which will form a permanent repair. Note that it will be necessary to remove a protective covering from the top surface of the patch after it has sealed in position. Inner tubes made from synthetic rubber may require a special type of patch and adhesive, if a satisfactory bond is to be achieved.

7   Before replacing the tyre, check the inside to make sure the agent that caused the puncture is not trapped. Check also the outside of the tyre, particularly the tread area, to make sure nothing is trapped that may cause a further puncture.

8   If the inner tube has been patched on a number of past occasions, or if there is a tear or large hole, it is preferable to discard it and fit a replacement. Sudden deflation may cause an accident.

9   To replace the tyre, inflate the inner tube sufficiently for it to assume a circular shape but only just. Then push it into the tyre so that it is enclosed completely. Lay the tyre on the wheel at an angle and insert

Tyre changing sequence - tubed tyres

 A Deflate tyre. After pushing tyre beads away from rim flanges push tyre bead into well of rim at point opposite valve. Insert tyre lever adjacent to valve and work bead over edge of rim.

 B Use two levers to work bead over edge of rim. Note use of rim protectors

C Remove inner tube from tyre

 **D** When first bead is clear, remove tyre as shown

 E When fitting, partially inflate inner tube and insert in tyre

F Work first bead over rim and feed valve through hole in rim. Partially screw on retaining nut to hold valve in place.

 G Check that inner tube is positioned correctly and work second bead over rim using tyre levers. Start at a point opposite valve.

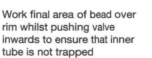 **H** Work final area of bead over rim whilst pushing valve inwards to ensure that inner tube is not trapped

the valve through the hole in the wheel rim. Attach the locking cap on the first few threads, sufficient to hold the valve captive in its correct location.

10 Starting at the point furthest from the valve, push the tyre bead over the edge of the wheel rim until it is located in the central well. Continue to work around the tyre in this fashion until the whole of one side of the tyre is on the rim. It may be necessary to use a tyre lever during the final stages.

11 Make sure there is no pull on the tyre valve and again commencing with the area furthest from the valve, ease the other bead of the tyre over the edge of the rim. Finish with the area close to the valve, pushing the valve up into the tyre until the locking cap touches the rim. This will ensure the inner tube is not trapped when the last section of the bead is edged over the rim with a tyre lever.

12 Check that the inner tube is not trapped at any point. Reinflate the inner tube, and check that the tyre is seating correctly around the wheel rim. There should be a thin rib moulded around the wall of the tyre on both sides, which should be equidistant from the wheel rim at all points. If the tyre is unevenly located on the rim, try bouncing the wheel when the tyre is at the recommended pressure. It is probable that one of the beads has not pulled clear of the centre well.

13 Always run the tyres at the recommended pressures and never under or over-inflate. The correct pressures for solo use are given in the Specifications Section of this Chapter. It should be remembered that the small size of the tyres means that the loss of a small quantity of air will result in a significant drop in pressure, so regular checks on tyre pressure should not be overlooked.

14 Tyre replacement is aided by dusting the side walls, particularly in the vicinity of the beads, with a liberal coating of French chalk. Washing-up liquid can also be used to good effect, but this has the disadvantage of causing the inner surfaces of the wheel to rust.

15 Never fit a tyre that has a damaged tread or side wall. Apart from the legal aspects, there is a very great risk of a blow-out which can have serious consequences on any two wheel vehicle.

16 Tyre valves rarely give trouble, but it is always advisable to check whether the valve itself is leaking before removing the tyre. Do not forget to fit the dust cap, which forms an effective second seal. The tyre valve dust cap is often left off when a tyre has been replaced, despite the fact that it serves an important two-fold function. Firstly, it prevents dirt or other foreign matter from entering the valve and causing the valve to stick open when the tyre pump is next applied. Secondly, it forms an effective second seal so that in the event of the tyre valve leaking, air will not be lost.

12 Valve cores and caps

1 Valve cores seldom give trouble, but do not last indefinitely. Dirt under the seating will cause a puzzling 'slow-puncture'. Check that they are not leaking by applying spittle to the end of the valve and watching for air bubbles.

2 A valve cap is a safety device, and should always be fitted. Apart from keeping dirt out of the valve, it provides a second seal in case of valve failure, and may prevent an accident resulting from sudden deflation.

Chapter 6 Electrical system

Refer to Chapter 7 for information relating to the SA50 Met-in

Contents

General description 1
Testing the electrical system: general information 2
Wiring: layout and examination 3
Battery: examination and maintenance 4
Charging system: checking the output 5
Regulator/rectifier: testing 6
Flywheel generator coils: testing 7
Lighting resistor: testing 8
Fuse: location and renewal 9
Electric starter system: fault diagnosis 10
Electric starter system: checking the relay 11

Electric starter system: checking the starter motor 12
Switches: examination and testing 13
Oil level sensor: removal and testing 14
Fuel gauge and sensor: removal and testing 15
Instrument panel warning bulbs: renewal 16
Headlamp: bulb renewal and beam alignment 17
Turn signal lamps: bulb renewal 18
Turn signal relay: location and renewal 19
Horn: location and renewal 20
Stop/tail lamp: bulb renewal 21

Specifications

Electrical system
Type	Flywheel generator, direct lighting
Voltage	12V
Earth	Negative (−)

Generator
Ouput	12V 114W @ 5000 rpm
Charging coil resistance (white to earth)	0.2 - 0.9 ohms
Lighting coil resistance (yellow to earth)	0.1 - 0.8 ohms

Battery
Capacity	12 volt, 3 Ah
Charge rate:	
Standard	0.3A
Fast charge	3.0A
Charge time:	
Standard	5 hours
Fast charge	30 minutes

Fuse
Rating	7A

Bulbs (all 12 volt)
Headlamp	25/25W
Stop/tail	5/21W
Turn signals	15W (21W later models)
Instrument light	1.7W x 2
High beam warning	1.7W
Turn signal warning	3.4W

1 General description

The Honda Vision models are equipped with a 12 volt negative earth electrical system powered from the flywheel generator. The headlamp and tail lamp circuit runs direct from a lighting coil on the generator stator using the alternating current (ac) output without regulation or rectification. The rest of the system, including the turn signals, horn, brake lamp and instruments use direct current (dc). This is derived from a separate charging coil, and passes through an electronic regulator/rectifier unit before being passed to the battery.

The battery is rated at 12 volt 3 Ampere hour (Ah) and is of sealed construction. It provides a reserve of power while the engine is idling, and thus not charging, and also allows a stable supply for circuits such as the turn signals, which would be adversely affected by one that fluctuates. It also provides a reserve of power to operate the electric starter.

2 Testing the electrical system: general information

1 The electrical system lends itself to fairly comprehensive testing of its component parts. A certain amount of preliminary dismantling is necessary to gain access to the components to be tested and full information on removing and fitting the various body panels mentioned in the following Sections of this Chapter is given in Chapter 4 of this Manual.

2 Simple continuity checks may be made using a dry battery and bulb arrangement, but for most of the tests in this Chapter a pocket multimeter can be considered essential. Many owners will already possess one of these devices, but if necessary they can be obtained from eletrical specialists or mail order companies.

3 Care must be taken when performing any electrical test, because some of the electronic assemblies can be destroyed if they are connected incorrectly or inadvertently shorted to earth. Instructions regarding meter probe connections are given for each test, and these should be read carefully to prevent any accidental damage during the test. Note that separate amp, volt and ohm meters may be used in place of the multimeter if necessary, having the appropriate test ranges.

4 Where test equipment is not available, or the owner feels unsure of the procedure described, it is recommended that professional assistance is sought. Do not forget that a simple error can destroy a component such as the rectifier, resulting in expensive replacements.

5 To gain access for most of the work described in this Chapter it will be necessary to remove one or more of the body panels, most notably the side panels, to reach the majority of the electrical components; the front panel to gain access to the multi-pin harness connector block; and the headlamp nacelle sections to reach the switches, headlamp and turn signal lamps, and the instrument panel. The procedure for removing the various panels is described in Chapter 4, Section 7.

3 Wiring: layout and examination

1 The wiring harness is colour coded and will correspond with the wiring diagram at the back of this manual. Where socket connectors are used they are designed so that reconnection can be made only in the one correct position.

2 Visual inspection will show whether any breaks or frayed outer coverings are giving rise to short circuits. Another source of trouble may be the snap connectors and sockets, where the connector has not been pushed home fully in the outer housing.

3 Intermittent short circuits can often be traced to a chafed wire that passes through or is close to a metal component, such as a frame member. Avoid tight bends in the wire or situations where the wire can become trapped between casings.

4 Note that the main harness passes through a large connector block located behind the front panel. This is a convenient point to start when testing wiring runs and circuits. It is a good idea to mark the various connector strips with numbers to indicate their correct relative position and direction before disconnecting them.

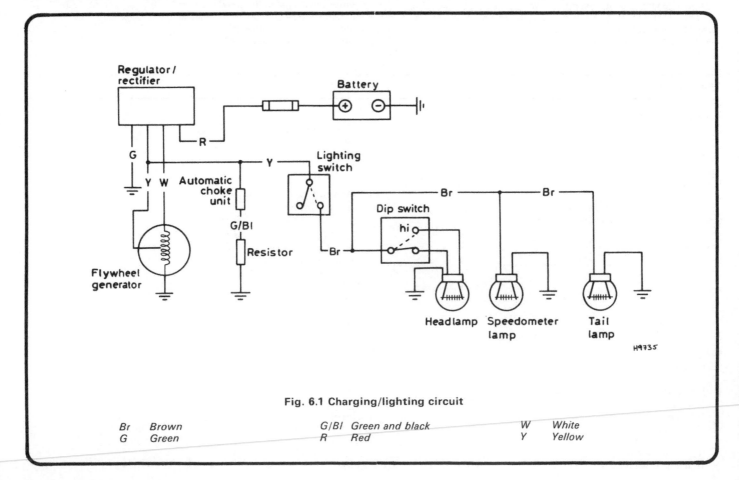

Fig. 6.1 Charging/lighting circuit

Br	Brown	G/Bl Green and black	W White
G	Green	R Red	Y Yellow

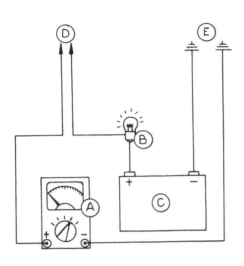

Fig. 6.2 Simple continuity test circuits

A	Multimeter	D	Positive probe
B	Bulb	E	Negative probe
C	Battery		

3.4a The main connector block (arrowed) is located behind the front panel

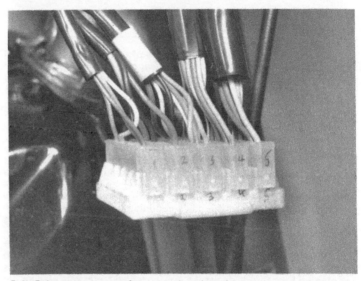

3.4b Release connector from panel and mark connector strips before removing them

4 Battery: examination and maintenance

1   The Vision models are fitted with a 12 volt 3 Ah battery of sealed construction. This means that the usual inspection and topping up procedures are not required (or possible). To gain access to the battery, unlock and open the seat, then lift the battery compartment cover. When removing the battery, disconnect the negative (green) lead first, then the positive (red) lead. When installing the battery, apply petroleum jelly to the terminals to prevent corrosion. Fit the positive lead first, followed by the negative lead.

2   The sealed construction of the battery precludes normal methods of testing using a hydrometer. Honda recommend that the battery is tested using a digital multimeter, Part Number 07411-0020000. A normal moving coil meter may not be sufficiently accurate or sensitive to give a definitive assessment of the battery condition, so it is probably worth asking a Honda dealer or a battery supplier to carry out the test for you. A fully charged battery in good condition should show a reading of 13.0 – 13.2 volts, whilst a reading of less then 12.3 volts indicates the need for charging. A battery which is unable to hold a charge, or which does not reach the fully charged voltage, even when charged off the machine, will require renewal. Again, have the battery checked professionally before ordering a new one.

3   During the winter months, or where numerous short journeys are undertaken, it is likely that the battery will require external charging from time to time. This should be done with the battery removed from the machine. It is strongly recommended that a small trickle-charger of the type designed specifically for use on motorcycle batteries is used. The more common car-type chargers are voltage controlled, which means that it is not possible to regulate the charging current. It is likely that the charging current from equiment of this type will be much too high for small batteries, and there is a real risk of damage.

4   The sealed construction of the battery imposes limits on the rate at which gassing can be vented during charging. If too high a charge rate is applied, there is risk of pressure bursting the battery. Also, a high charge rate may overheat the battery, warping the plates and shortening its life significantly. The normal charge rate is $1/10$ of the rated capacity of the battery, in this case 0.3 amps. A fully discharged battery will require 5 hours of charging at this rate. In an emergency, a charge rate of up to 3.0 amps may be applied for a maximum of 30 minutes. Be warned that this is for emergencies only; it will shorten the life of the battery. If using the 'fast charge' rate, check the battery case frequently and stop charging at once if it gets more than slightly warm.

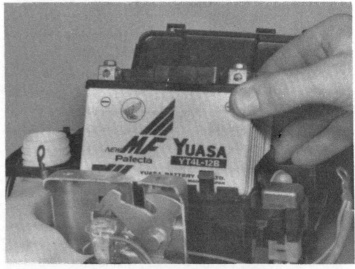

4.1 Battery is housed in compartment next to oil tank

5 Charging system: checking the output

1   This test requires the battery to be in good condition and fully charged – if in any doubt refer to Section 4. Also needed are an ammeter and a voltmeter. If you do not have access to this equipment or are unsure of how to use it, the test should be performed by a Honda dealer. The test can be performed after running the engine until normal operating temperature has been reached. Stop the engine and then open the seat; no body panels need be removed.

2   Open the fuse holder next to the battery and disconnect and separate the battery lead at the fuse terminal. Connect the ammeter between the lead and the fuse terminal. Connect the voltmeter across the battery terminals, observing polarity.

3   Start the engine and observe the two meters, while gradually increasing the engine speed. The ammeter should show a slight charge (no specific figures are available), whilst the volt meter should read 14 – 15 V. If the readings are other than specified, check first for loose or broken connections. If the ammeter shows a discharge, or if the voltage readings are abnormally high or low, a faulty regulator/rectifier is indicated. Check this by measuring the regulator/rectifier resistances, or better still, by fitting a new unit. If the fault persists, check the condition of the alternator.

6 Regulator/rectifier: testing

1   The regulator/rectifier takes the form of a sealed, finned alloy unit bolted to the frame immediately to the rear of the oil tank. Access to the unit requires the removal of the rack and both side panels (see Chapter 4, Section 7). The unit rectifies the output of the charging coil, and also controls the charge rate applied to the battery. It is a fairly robust unit, but may be damaged by accidental short circuits, or in time, by mechanical damage from road shocks. This makes it a prime suspect in the event of a charging system fault.

2   Honda recommend that one of two specified test meters are used to carry out the test; either the Sanwa Electical Tester SP-10D (Part Number 07308-0020000) set on the x K ohms range, or the Kowa Electrical Tester (TH-5H) set on the x 100 ohms range. If a meter of a different type is used, false readings may result, although in practice a reasonable indication of the unit's condition should still be given.

3   Unplug the wiring connector from the side of the unit then, using the accompanying table and line drawing, check the resistance between the various pairs of terminals. If any one reading is significantly outside the specified figure the unit can be assumed to have failed internally and must be renewed. Have your findings confirmed by a Honda dealer if a non-specified meter was used for the test.

7 Flywheel generator coils: testing

1   The condition of the generator stator coils can be checked without removing the stator. Start by removing the left-hand side panel to gain access to the wiring connections near the footboard. Trace the wiring back from the flywheel generator and disconnect it at the relevant connectors. Measure the resistance between the white lead and earth, and then the yellow lead and earth, comparing the readings obtained with those shown below.

White lead to earth .... 0.2 – 0.9 ohms
Yellow lead to earth .... 0.1 – 0.8 ohms

2   If the readings obtained are significantly outside those shown, the stator coil concerned is defective and must be renewed. Note that a reading of zero resistance indicates that the coil has shorted, whilst a reading of infinite resistance denotes an open circuit, or break in the wiring. Honda supply the stator as an assembly; individual coils cannot be supplied separately. The alternative to renewal of the entire stator is to ask an auto electrical specialist if the affected coil can be rewound. Failing this, try local motorcycle breakers who may be able to supply a good second-hand stator assembly.

8 Lighting resistor; testing

The headlamp circuit is stabilised by a small resistor mounted on the frame seat catch upright just to the rear of the battery casing. This should be checked if frequent bulb failure is experienced. Using a multimeter, check the resistance between the green/black lead and earth. A reading of 5.0 ohms should be indicated. If not, renew the resistor.

9 Fuse: location and renewal

1   The electrical system is protected by a 7 amp fuse in an in-line holder in the battery positive (red) lead. Access to the fuse is from below the seat; no body panels need be disturbed.

2   The fuse is fitted to protect the electrical system in the event of a short circuit or sudden surge. It is, in effect, an intentional 'weak link' which will blow in preference to the circuit burning out.

3   Before replacing a fuse that has blown, check that no obvious short circuit has occurred, otherwise the replacement fuse will blow immediately it is inserted. It is always wise to check the electrical circuit thoroughly, to trace the fault and eliminate it.

4   When a fuse blows while the machine is running and no spare is available, a 'get you home' remedy is to remove the blown fuse and wrap it in silver paper before replacing it in the fuseholder. The silver paper will restore the electrical continuity by bridging the broken fuse wire. This expedient should **never** be used if there is evidence of a short circuit or other major electrical fault, otherwise more serious damage will be caused. Replace the 'doctored' fuse at the earliest possible opportunity, to restore full circuit protection. It follows that spare fuses that are used should be replaced as soon as possible to prevent the above situation from arising.

10 Electric starter system: fault diagnosis

1   In the event of a failure of the starter circuit, always check first that the battery is in good conditon and fully charged. The motor draws a very heavy current compared with the rest of the electrical system, and it is not uncommon for the battery to become discharged just enough to prevent it from operating the starter successfully. If in doubt, remove the battery and recharge it as described in Section 4.

2   If the battery is fully charged and the starter still fails to operate, check the operation of the starter relay as described in Section 11, then, if the fault persists, check the motor itself.

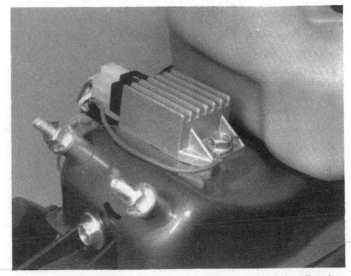

6.1 Regulator/rectifier is bolted to rear of frame, behind the oil tank

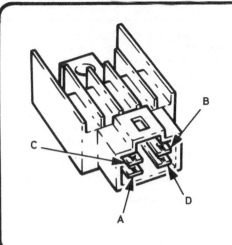

- \ +	A	B	C	D
A		∞	0.5–10K	∞
B	∞		∞	5–100K
C	∞	∞		∞
D	∞	5–100K	∞	

H.16832

Fig. 6.3 Regulator/rectifier unit test

8.1 Resistor is held by a single bolt to seat catch upright

9.1 (A) Fuse holder, (B) Starter relay

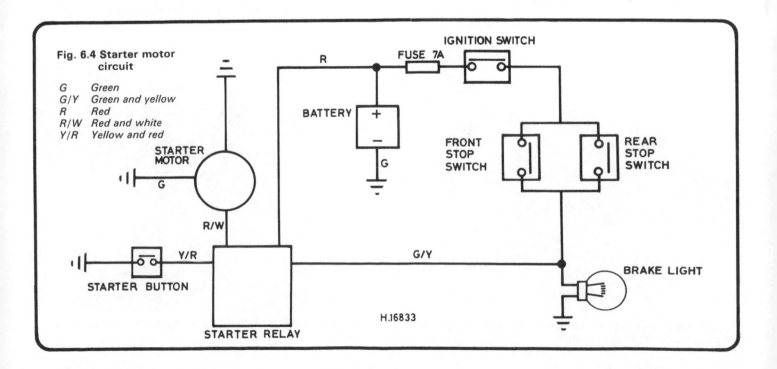

Fig. 6.4 Starter motor circuit

G   Green
G/Y Green and yellow
R   Red
R/W Red and white
Y/R Yellow and red

STARTER MOTOR

BATTERY

FUSE 7A

IGNITION SWITCH

FRONT STOP SWITCH

REAR STOP SWITCH

STARTER BUTTON

STARTER RELAY

BRAKE LIGHT

H.16833

11 Electric starter system: checking the relay

1   The starter relay acts as a remote switch, capable of switching the high current required by the starter motor. When the ignition is switched on, the rear brake locked on, and the starter button pressed, a small current is drawn by the relay windings as the main contacts are closed. These carry the high current drawn by the motor, and prevent arcing at the handlebar switch contacts. The operation of the relay can be checked using the machine's battery and a multimeter.

2   The relay takes the form of a sealed unit, held in a rubber mounting attached to the right-hand side of the battery housing, next to the main fuse. Access to the unit requires the removal of the rack and the right-hand side panel (see Chapter 4, Section 7).

3   Unplug the relay wiring and remove it from its holder. Make up two jumper leads, connecting one to the green/yellow lead terminal and its other end to the battery positive terminal. Connect the second jumper lead to the yellow/red terminal of the unit, leaving its other end unconnected at present.

4   Connect a battery and bulb continuity tester or a multimeter set on the resistance (ohms) range between the red terminal and the red/white terminal. No continuity should be shown at this stage, but the circuit should be completed when the free end of the second jumper lead is touched on the battery negative terminal. This should be accompanied by an audible click as the relay contacts close.

5   The above test should normally establish whether or not the relay is working. In rare cases the relay will work well enough to indicate a continuity reading, but the contacts may have become sufficiently burnt not to be able to cope with the heavy starter motor current. If this is the case it will be necessary to check by fitting a new unit.

12 Electric starter system: checking the starter motor

1   If the above check of the starter relay has indicated a fault in the motor, it should be removed for checking. Start by raising the seat, then isolate the battery by disconnecting the battery negative lead. Referring to Chapter 4, Section 7, remove both side panels to gain access to the motor. Working from the right-hand side of the machine, remove the two starter motor mounting bolts and pull the motor clear of the crankcase. Pull back the protective shroud and remove the single screw which secures the starter motor lead to the terminal. Disconnect the lead and lift the motor away for further testing.

2   The operation of the motor can be checked using two thick pieces of wire as jumper leads. Connect one wire between the battery positive terminal and the motor terminal, keeping it well away from the motor casing. Connect the second jumper to the motor casing, using a nut and bolt to attach it to one of the mounting lugs. Check that when the free end of the second jumper is touched on the battery negative terminal, the motor turns freely. When viewed from the spindle end, the motor should turn clockwise.

3   If the motor fails to work normally, little can be done to repair it because internal parts are not available separately. There is, however, nothing to be lost in attempting a repair. The commutator segments can be cleaned with very fine abrasive paper and an auto-electrical specialist may be able to find a brush set to suit; genuine brushes are not available from the manufacturer. If the brushes are removed some patience will be required to refit them in their holders. Failing this, it will be necessary to renew the motor, or to obtain a second-hand unit from a motorcycle breaker.

4   When refitting the motor, note that a new O-ring should be fitted to the nose of the motor casing. Grease the O-ring to facilitate installation, then reconnect the motor lead. Fit the terminal screw to hold the lead in place, and slide the protective sleeve over the terminal. Refit the motor, noting that the upper mounting bolt passes through the earth terminal. Tighten the bolts evenly and securely.

12.3a Starter motor casing is secured by two screws

12.3b Cover removed to reveal armature and brush assembly. **Note:** No parts are available for the motor

12.3c The brushplate removed, showing brushes. Installation is *very* awkward

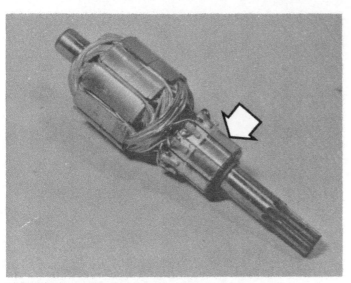

12.3d Check commutator segments (arrowed) for wear or discoloration. Clean with very fine abrasive paper

12.3e Check condition of O-ring before installing motor

13 Switches: examination and testing

1   The function of each switch can be easily tested using a multimeter or a battery and bulb arrangement to check continuity in each of the switch positions. The switch wiring is routed down the side of the steering column, where it terminates in a connector strip. Each strip plugs into a single large multi-pin connector. To gain access to the connector, remove the front panel as detailed in Chapter 4, Section 7. The connector is attached to the front of the legshield moulding and is secured by a single screw. Before disconnecting the various connector strips, it is a good idea to number each one so that it can be refitted in its original position.

2   Referring to the wiring diagram at the end of this Chapter, note that each switch is represented in diagrammatic form, showing the wiring colours and their corresponding switch positions. Identify the connector strip belonging to the switch in question and unplug it. Connect the multimeter or battery and bulb arrangement across each pair of leads to be tested and check that it functions as shown on the switch diagram when the switch is operated. If a fault is indicated, or if operation is erratic, the switch must either be repaired or renewed as described below.

Ignition switch

3   The ignition switch is combined with the steering lock, the assembly being bolted to the side of the steering head. The switch can be reached after removing the front panel and legshield as detailed in Section 7 of Chapter 4. If the operation of the switch has been affected by corrosion or water, try soaking it in WD40 and operating the switch repeatedly. If this fails to effect a cure, the switch will have to be renewed, together with the steering lock. If a new switch is fitted, remember that the new key will not necessarily fit the remaining locks on the machine, so either these will have to be renewed at the same time, or two keys will have to be carried.

Handlebar switches

4   The handlebar switches are positioned at the ends of the handlebar nacelle rear section, and may be reached after removing the rear section as detailed in Section 7 of Chapter 4. If the operation of a switch has been affected by corrosion or water, try soaking it in WD40 and operating the switch repeatedly. This is quite often successful, the back of each switch being so designed that the fluid will penetrate readily. If this fails to effect a cure, the switch will have to be renewed. Before ordering a replacement, however, you may wish to attempt to dismantle and physically clean the switch terminals. The method of doing so is self-evident, but be warned that the switch may tend to fly apart when disturbed! It is suggested that the switch is removed and the work carried out on a clean bench so that small parts such as springs are not mislaid.

Brake lamp switches

5   The two brake lamp switches are small plunger-type units mounted on the brake lever stocks. The switches are visible on the underside of the lever assembly, but if removal is necessary or access to the wiring connectors is required, they may be reached after removing the rear section of the handlebar nacelle as detailed in Section 7 of Chapter 4. If the operation of a switch has been affected by corrosion or water, try soaking it in WD40 and operating the switch repeatedly. If this fails to restore operation, renew the switch; no dismantling is possible.

13.3 Ignition switch is retained by two mounting bolts

13.4a Handy access hole allows WD40 to be sprayed onto contacts without disturbing switch

13.4b Switch clusters can be removed if required ...

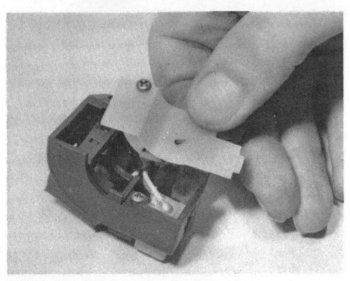

13.4c ... as can covers on underside of each switch

13.4d Individual switches are removed at your peril; beware of escaping springs and detent balls

14 Oil level sensor: removal and testing

1   The oil level sensor consists of a rubber-encased float switch mounted in the top of the oil tank. It is a simple and robust unit, and is unlikely to give trouble. To gain access to the sensor for testing, remove the rack and the side panels as described in Chapter 4, Section 7. Unplug the sensor leads from the top of the unit, and pull it out of the oil tank.

2   Using a multimeter or a battery and bulb continuity tester, check that the sensor conducts when the float is in the fully down position, and is isolated when the float is raised. If all seems well, reconnect the sensor leads and switch on the ignition. Try raising and lowering the float and note the effect on the warning lamp. If the problem persists, check the wiring for breaks or loose connections, and make sure that the bulb in the warning lamp is intact and making good contact.

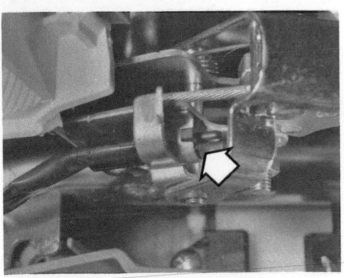

13.5 Brake lamp switches are mounted near lever pivots

15 Fuel gauge and sensor: removal and testing

1   The fuel gauge sensor is mounted in the top of the fuel tank and can be removed for testing after the seat has been opened. Slide back the plastic cover which covers the terminals on the unit to reveal the

retaining ring. This can be released by gripping the indentations in its edge with a pair of slip-jointed pliers. Turn the retainer anticlockwise until it comes free, then slide it clear of the unit. The sensor itself can now be lifted out of the tank, noting that it must be manoeuvred very carefully to avoid damage to the float or float arm.

2   The sensor consists of a float-operated variable resistance, and this can be checked by measuring the resistance of the unit at various float positions. Connect a multimeter across the terminals, setting it to the resistance (ohms) range. The resistance with the float fully raised (full position) should be 9.4 – 10.0 ohms, whilst at the fully lowered positon (empty) it should read 90 – 100 ohms. The meter needle should move smoothly and progressively between the two readings as the float arm is raised and lowered. If the readings are outside those given, or the resistance is erratic, the problem is usually corrosion of the resistance windings or the wiper arm.

3   Although a unit showing these symptoms is theoretically in need of renewal, it is possible to remove the pressed steel cover to gain access to the windings and wiper arm, or moving contact. The electrical contact surfaces can be cleaned using very fine abrasive paper or similar; an ink eraser is ideal for this purpose and will not cause severe scratching. Finally, spray the windings with WD40 or similar, then repeat the resistance check to see if normal operation has been restored.

4   If the sensor operates normally, but there is still a fault, check that the electrical system in general functions normally by operating the turn signals, then connect the sensor leads without refitting the unit in the tank. Turn on the ignition switch and operate the float manually. If the gauge fails to respond, check back through the wiring to eliminate breaks or short circuits. If this does not resolve the problem, renew the gauge mechanism.

5   The gauge movement is incorporated in the speedometer head, access to which requires the removal of the rear section of the handlebar nacelle (see chapter 4, Section 7). Remove the speedometer from the moulding, then separate the speedometer head from the plastic lens (three screws) and remove the fuel gauge mechanism. Reassemble the new unit by reversing the removal sequence.

6   Before refitting the sender unit in the tank, check the sealing ring and renew it if necessary. When refitting the sender unit, take care not to bend the float or float arm. Align the tank's tabs with the sender unit cutouts to fully insert it, then twist the sender unit away from the tabs to secure it (the correct position is when the arrows on the sender and tank are in alignment – see photo 15.1a).

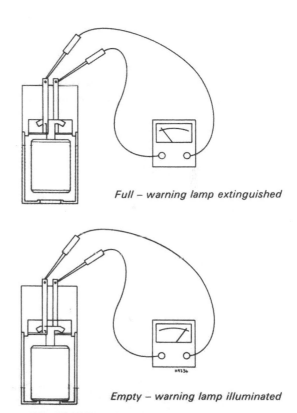

Full – warning lamp extinguished

Empty – warning lamp illuminated

Fig. 6.5 Oil level sensor test positions

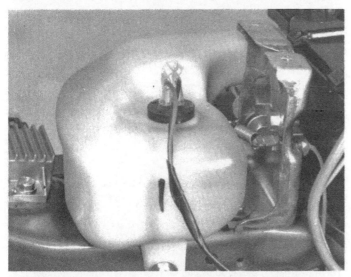

14.1 Oil level sensor can be removed from tank for checking

15.1a Pull back plastic cover and release retainer ring ...

15.1b ... by twisting it with slip-joint pliers

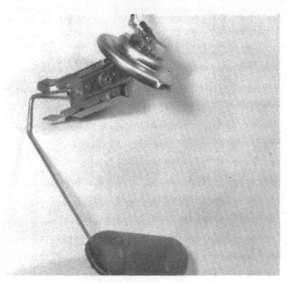

15.1c Fuel gauge sensor can now be manoeuvred out of tank

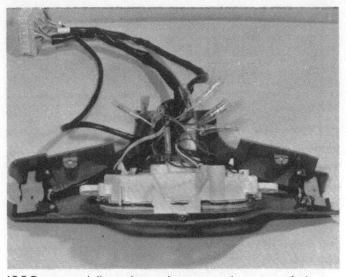

15.5 Remove and dismantle speedometer to gain access to fuel gauge head

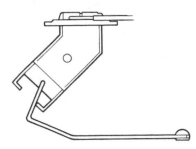

Full position

16.1a Instrument panel bulbs can be reached after removing rear section of headlamp nacelle

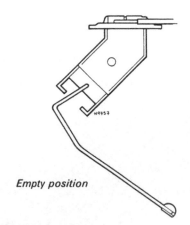

Empty position

Fig. 6.6 Fuel gauge sender testing positions

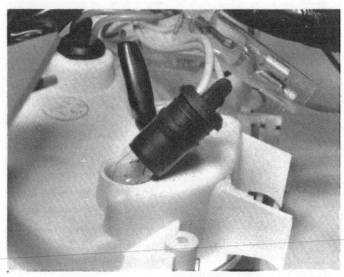

16.1b Bulb holders are a push-fit in back of speedometer

16 Instrument panel warning bulbs: renewal

Access to the instrument panel bulbs requires the removal of the rear section of the handlebar nacelle. This is described in detail in Section 7 of Chapter 4. Each of the bulbs is housed in a rubber bulbholder, these being a push fit in the back of the instrument panel casing. Identify the bulb in question and withdraw its bulbholder. The bulb is of the capless type and is a push-fit in its holder. When fitting a new bulb, ensure that it is of the correct wattage.

Fig. 6.7 Instrument panel

1   Top cover
2   Screws
3   Instrument panel
4   Bulbs and wiring
5   Speedometer
6   Fuel gauge
7   Screw
8   Housing
9   Warning lamp wiring
10 Bulb – 2 off
11 Bulb
12 Screw – 2 off
13 Screw
14 Screw – 3 off

H.16834

17 Headlamp: bulb renewal and beam alignment

1   Access to the headlamp and parking lamp bulbs requires the removal of the rear section of the handlebar necelle. This is described in detail in Section 7 of Chapter 4. To remove the headlamp bulb, depress the bulbholder and turn it anticlockwise. Lift the bulbholder clear to reveal the back of the bulb. The new bulb can be fitted by reversing the above sequence.

2   If the headlamp requires adjustment, this can be carried out by moving the adjuster screw to the required position. The screw is located in the underside of the headlamp rim. Note that most countries have laws governing the setting of beam alignment, and adjustments must be made with this in mind. In the UK, the headlamp beam must be set so that the light will not dazzle a person standing at a distance greater than 25 feet from the lamp, and whose eye level is not less than 3 feet 6 inches above that plane.

3   This can be approximated by measuring the height of the centre of the lamp from the ground. Find a level surface with a flat wall or similar at one end, and make a mark at the measured height on the wall. With the machine on its wheels and with the rider seated, check that the most concentrated part of the beam aligns with the mark on the wall.

17.1a Release headlamp bulbholder assembly by twisting retainer ...

17.1b ...then lift bulb out of the reflector

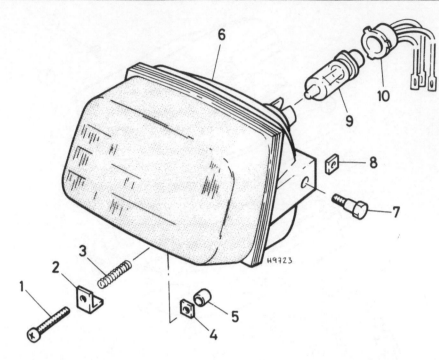

Fig. 6.8 Headlamp

1 Beam height adjusting
   screw
2 Plate
3 Spring
4 Nut
5 Cover
6 Reflector and glass
7 Bolt – 2 off
8 Nut – 2 off
9 Bulb
10 Bulbholder

18 Turn signal lamps: bulb removal

1 Bulb failure is by far the most common cause of turn signal problems, and is characterised by one lamp failing to flash, while the other flashes dimly and rapidly. The other pair of lamps will be unaffected. Having identified the bulb at fault, proceed as follows.

2 In the case of early model front turn signal lamps, remove the screw at the outer edge of the rear section of the handlebar nacelle on the side affected. Now remove the single screw which secures the lens and shroud, lifting them away to reveal the bulb. On later models the lamps are mounted on small brackets which extend from the legshield. A single screw retains each lens.

3 On early models the rear turn signal lamps are incorporated in the rear lamp cluster and may be reached after the lens unit has been released by removing its retaining screws. On later models the lamps are mounted on a bar projecting from the rear carrier. A single screw retains each lens.

4 In each case, fit a new bulb, and check that the turn signals operate normally before refitting the lens. If the fault persists or the bulb proves not to have failed, check for a wiring fault in the turn signal circuit. Note that failure of one side of the circuit only is not normally attributable to the turn signal relay.

19 Turn signal relay: location and renewal

1 If there is a fault in the turn signal circuit, always check the bulbs (see Section 18), wiring and switch before turning attention to the relay. A partially discharged battery will also affect the operation of the turn signal system. A fault in the relay will affect all four lamps; not just one lamp or pair of lamps.

2 If the relay develops a fault it will be necessary to renew it. Access to the relay requires the removal of the front panel as described in Chapter 4, Section 7. The relay is mounted to the left of the horn, when viewed from the front of the machine.

3 Renewal is a simple matter of unplugging the defective relay and plugging in a new one. Check that the system functions normally, then refit the front panel.

20 Horn: location and renewal

The horn is mounted on the front of the machine, beneath the front panel. This must be removed to gain access to the horn, following the

instructions given in Chapter 4, Section 7. Check the horn by removing it and connecting a 12 volt battery to its terminals. If it fails to sound it must be renewed; repair is not practicable. If the horn operates under test, but not on the machine, check the wiring and switch operation as described earlier in this Chapter.

21 Stop/tail lamp: bulb renewal

Access to the stop/tail bulb is gained by removing the two screws which retain the lens unit and lifting it away. When renewing the bulb, check that it is of the correct wattage, and note that the pins are offset to prevent it from being fitted incorrectly. If the brake or tail lamp fails to operate despite the bulb being intact, trace back through the wiring, checking for a break or short, and check the operation of the lighting and brake lamp switches.

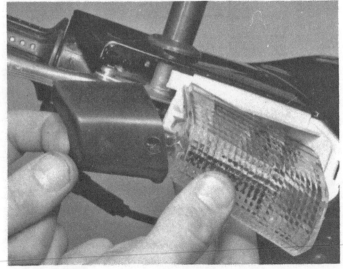

18.2a Remove screws to free lens and plastic shroud ...

18.2b ... to reveal front turn signal bulb

19.2 Turn signal relay is mounted behind front panel, next to the horn

20.1 Horn is attached to flexible mounting strip behind front panel

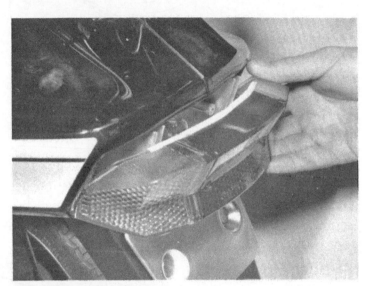

21.1a Release the rear lamp lens to gain access to stop/tail lamp and rear turn signal bulbs (early models)

21.1b Stop/tail lamp bulb has offset bayonet fitting

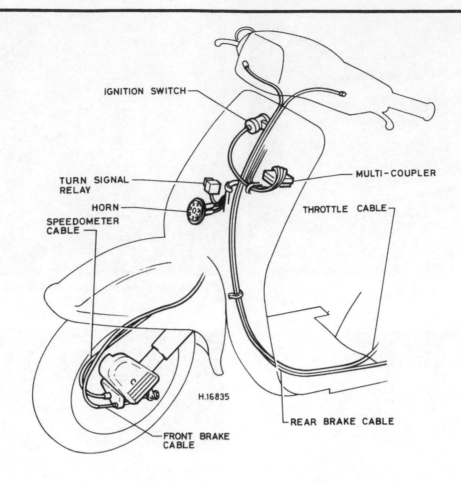

IGNITION SWITCH

TURN SIGNAL
RELAY

HORN

SPEEDOMETER
CABLE

MULTI-COUPLER

THROTTLE CABLE

REAR BRAKE CABLE

FRONT BRAKE
CABLE

H.16835

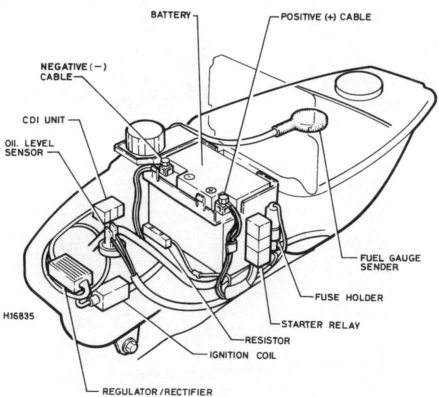

BATTERY

POSITIVE (+) CABLE

NEGATIVE (−)
CABLE

CDI UNIT

OIL LEVEL
SENSOR

FUEL GAUGE
SENDER

FUSE HOLDER

STARTER RELAY

RESISTOR

IGNITION COIL

REGULATOR /RECTIFIER

H.16835

Fig. 6.9 Electrical and ignition component locations

The SA50-M Vision Met-in Model

Chapter 7 The SA50 Vision Met-in

Contents

Introduction 1
Routine maintenance: schedule 2
Routine maintenance: operations 3
Engine and lubrication system: modifications....................... 4
Ignition components: location and testing............................. 5
Body panels: removal and refitting.. 6
Battery: removal and refitting 7
Regulator/rectifier unit: location and testing........................ 8

Flywheel generator coils: testing.. 9
Starter relay: location 10
Resistors: location and testing 11
Fuel gauge sender unit: removal, testing and refitting 12
Headlamp: removal and refitting... 13
Turn signals: general... 14
Instrument panel: removal and refitting................................ 15
Handlebar switches: removal and refitting............................ 16

Specifications

Model dimensions and weights

Overall length	1710 mm (67.3 in)
Overall width	650 mm (25.6 in)
Overall height	1025 mm (40.4 in)
Wheelbase	1200 mm (47.2 in)
Ground clearance	85 mm (3.3 in)
Seat height	Not available
Dry weight	66 kg (146 lbs)

Specifications relating to Chapter 1

Engine

Compression ratio	6.5:1
Maximum power	Not available
Maximum torque	Not available

Cylinder barrel

Diameter:	
With A mark	41.000 – 41.005 mm (1.6142 – 1.6144 in)
With no mark	41.005 – 41.010 mm (1.6144 – 1.6146 in)
Service limit	41.900 mm (1.6500 in)

Piston and rings

Piston diameter (4 mm from bottom of skirt)	40.955 – 40.965 mm (1.6124 – 1.6128 in)
Service limit	40.900 mm (1.6102 in)
Gudgeon pin bore ID	10.002 – 10.008 mm (0.3938 – 0.3940 in)
Service limit	10.03 mm (0.3949 in)

Transmission

Transmission ratio range	2.4:1 to 1.1:1
Drive belt width	15.5 mm (0.61 in)
Service limit	14.5 mm (0.57 in)
Driven (rear) pulley:	
Return spring free length	98.1 mm (3.862 in)
Service limit	92.8 mm (3.654 in)

Torque wrench settings

	kgf m	lbf ft
Cylinder head bolt	1.0	7.2
Flywheel rotor nut	3.8	27.5
Drive pulley nut	3.8	27.5
Inlet stub bolt	Not available	Not available
Carburettor bolt	Not available	Not available
Clutch centre nut	3.8	27.5
Clutch outer nut	3.8	27.5
Drive pulley cover bolts	0.45	3.25
Engine mounting bolt	3.5	27.5
Rear suspension unit lower mounting bolt	2.5	18.0

Specifications relating to Chapter 2

Fuel tank
Capacity.. 5.5 litres (1.2 Imp gal)

Carburettor
Identification No.. PA35A-A
Main jet.. 72
Slow jet... 35
Jet needle clip position.. 2nd groove from top

Torque wrench settings	kgf m	lbf ft
Fuel tap nut..	2.3	16.6

Specifications relating to Chapter 3

Ignition source coil
Resistance (black/red wire to earth) .... 500 – 900 ohms

Ignition timing.. 17° BTDC @ 1800 rpm ± 100 rpm

Specifications relating to Chapter 4

Front forks
Travel... 70 mm (2.75 in) (bottom link)
Spring free length .... 207 mm (8.15 in)
Service limit... 201 mm (7.91 in)

Rear suspension
Travel... 70 mm (2.75 in)
Spring free length .... 231 mm (9.09 in)
Service limit... 224.5 mm (8.84 in)

Torque wrench settings	kgf m	lbf ft
Steering stem top nut..	10.0	72.3
Steering stem locking ring...	0.9	6.5
Front wheel spindle nut..	4.5	32.5
Fork to trailing link pivot bolt*..	2.0	14.5
Suspension unit:		
Upper mounting bolt...	2.7	19.5
Lower mounting bolt...	1.8	13.0
Lower mounting nut..	0.9	6.5
Damper unit locknut...	Not available	Not available
Rear wheel stub axle nut..	11.0	79.5
Rear suspension unit:		
Upper mounting nut..	4.0	28.9
Lower mounting bolt...	2.5	18.0
Damper unit locknut*..	2.0	14.5

Use thread-locking compound on the threads of these fasteners

Specifications relating to Chapter 5

Brakes
Brake lining thickness (front and rear)................................. 3.0 mm (0.118 in)
Service limit... 1.5 mm (0.059 in)
Brake drum diameter (front and rear) .... 95 mm (3.74 in)
Service limit... 95.5 mm (3.76 in)

Tyres
Size (front and rear) 3.00-10-4PR

Torque wrench settings	kgf m	lbf ft
Front wheel spindle nut..	4.5	32.5
Rear wheel stub axle nut..	11.0	79.5
Brake arm bolt (front and rear)..	0.6	4.3

Specifications relating to Chapter 6

Generator
Output.. 12V 96W @ 5000 rpm
Charging coil resistance (white to earth)............................. 1.0 ohm

Battery
Charge rate:
Standard 0.4A
Fast charge.. 4.0A

Fuse
Rating... 10A

1 Introduction

1   This Chapter covers the Vision Met-in, introduced in October 1988. As the Met-in shares almost all components with the previous Vision models, it will be necessary to refer to the previous Chapters of this Manual for the majority of tasks. Where differences do occur, they will be found in this Chapter. It is therefore necessary to refer first to this Chapter, then if the information required cannot be found, it can be assumed that the operation will be identical to that given in the previous Chapters.

2   Cosmetically, the Met-in is quite different from the earlier Visions due to the fitting of more streamlined body panels and a revised headlamp and instrument nacelle. Additionally, the Met-in has a large storage compartment under the seat, suitable for holding a helmet.

3   Mechanical changes are few and result only in detail changes to specifications.

4   Although three versions of the Met-in have been released in the UK, changes have been limited to colour and graphics. Despite this, it is important to identify your model exactly when ordering spare parts. The following table gives details of the engine and frame numbers of these models and their dates of availability. Note that it is not sufficient to rely on the date of registration alone for identification because this does not always indicate the model's actual production date.

5   Frame and engine number locations are unchanged on the Met-in, but due to the body panel design, the frame number is now viewed by prising out the small inspection cover on the left-hand frame lower cover.

Model	Initial engine no.	Frame nos.	Dates of availability
SA50-J	AF05E-4600001	AF22-4600020 to 4640700	Oct '88-Nov '92
SA50-M	As above	AF22-4639948 on	Sept '91-Feb '93
SA50-P	As above	AF22-4700001 on	Dec '92-Apr '93
SA50-R	As above	As above	Apr '93-Nov '95

2 Routine maintenance: schedule

Note: *the following maintenance schedule is based on a mileage/time system, whichever comes first. Unless mentioned in other Sections of this Chapter, the tasks will be unchanged from that given in Routine maintenance at the beginning of this Manual. Many of the procedures require the removal of one or more body panels, details of which are given in Section 6 of this Chapter.*

Pre-ride checks
   Check the fuel level
   Check the engine oil level
   Check the operation of both brakes
   Check the tyre pressures and tyre condition
   Check that the throttle control operates correctly
   Check that the lights, turn signals, brake stop lamp, horn and speedometer all work correctly
   Check that the engine stop switch operates correctly

Every 1000 miles (1600 km)
   Renew the spark plug

Six monthly or every 2500 miles (4000 km)
   Check the fuel pipe for deterioration and leakage
   Check the throttle for smooth operation and check the cable free play
   Clean the air filter element
   Check and bleed the engine lubrication system
   Check the engine idle speed
   Check the operation of the front and rear brakes. Also check the extent of shoe wear
   Check the operation of the brake stop lamp switches
   Check the headlamp beam alignment
   Check the operation of the suspension
   Check the tyre pressures and check the tyres for wear and damage
   Check the wheels for damage

Every 1875 miles (1600 km)
   Decarbonise the cylinder head, barrel ports and exhaust system

Yearly or every 5000 miles (8000 km)
   Check the tightness of all accessible nuts, bolts and screws on the machine. Tighten to the specified torque wrench settings (where available)
   Check the control cables for wear and lubricate them
   Check the transmission for wear

18-monthly or every 7500 miles (12 000 km)
   Check the steering head bearing adjustment

3 Routine maintenance: operations

Spark plug inspection

1   Gain access to the spark plug via the inspection aperture at the bottom of the under-seat storage compartment. A slotted fastener retains the cover in place, which need only be slackened to allow the cover to be moved to one side.

2   Pull the suppressor cap from the top of the spark plug and unscrew the plug for inspection as described in Routine maintenance at the beginning of this Manual.

Oil pump cable adjustment

3   The oil pump cable setting is important to ensure the correct flow of oil to the engine. The cable from the throttle twistgrip splits into two mid-way along its length; one cable runs to the carburettor throttle slide and the other to the oil pump.

4   Prior to checking the oil pump adjustment, check for the correct amount of play at the throttle twistgrip. There should be between 2 and 6 mm freeplay, measured in terms of twistgrip rotation, before resistance is felt; if adjustment is necessary make this at the in-line adjuster at the handlebar end of the cable.

5   To gain access to the oil pump, remove the frame centre cover as described later in this Chapter. Have an assistant hold the throttle twistgrip fully open and observe the lug on the pump lever and the cutout in the pump body; if correctly set up, the two should align

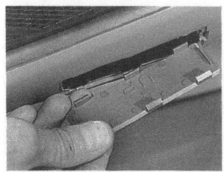

1.5 Remove inspection cover in frame lower cover to view frame number. Arrow on inside face indicates correct fitted position

3.1 Spark plug is accessible through inspection aperture in storage compartment

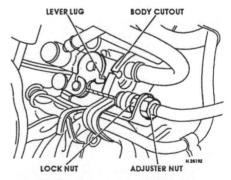

Fig. 7.1 Oil pump lever lug and body cutout should align with throttle fully open

exactly. If they don't align, slacken the locknut on the pump cable bracket and rotate the adjuster nut to align the marks. Tighten the locknut and recheck the setting. Refit the frame centre cover.

---

4   Engine and lubrication system: modifications

Piston/cylinder barrel selection

1   If examination and measurement of these parts necessitates their renewal, it is important to specify the cylinder/piston identification letter when ordering new parts.
2   Look for an identification letter on the exhaust port side of the barrel-to-crankcase gasket face (note that some barrels will have no mark at all). Similarly, inspect the piston crown, on the inlet side of the piston for similar marking.
3   Either no mark at all, or the letters A or B will be found. Quote this identification when ordering a new component.

Oil pump – removal and refitting

4   Remove the frame centre section as described later in this Chapter. Slacken the locknut of cable adjuster and slip the cable out of its bracket, disconnecting the cable trunnion from the oil pump lever.
5   Clamp the line from the oil tank and the line to the inlet adaptor to prevent oil loss and the entry of dirt, then pull both pipes off their stubs on the oil pump. Free the generator wire from the clamp on the pump bolt and remove the oil pump retaining bolt. Pull the pump out of the crankcase and recover the O-ring.
6   If the pump fails to operate or if there is severe wear on its drive gear teeth it must be renewed; no replacement parts are available.
7   Fit a new O-ring to the pump and apply molybdenum disulphide grease to the gear teeth. Slide the pump into the crankcase, making sure that its drive gear engages correctly. Make sure that the retaining bolt nut is held in the casting cutout, then refit the bolt and tighten it securely. Prime the oil pipes with oil (see Section 18 of Chapter 2), refit them on their stubs and secure with the retaining clips.
8   Reconnect the oil pump cable and carry out adjustment as described in Section 3. Check that there are no oil leaks and refit the frame centre cover.

---

5   Ignition components: location and testing

CDI unit

1   The CDI unit is located above the battery, next to the starter relay. Unlike the unit fitted to previous models, its internal resistances cannot be measured to determine its condition. The only performance test specified by the manufacturer, is on a spark gap tester, only normally available to a Honda dealer. Alternatively, the unit can be checked by the substitution of a known good unit.
2   If an ignition fault occurs, it is suggested that all other components in the ignition system are checked and eliminated first. Start the check in a logical manner, commencing first with the most likely items as described below.

(a)   *Substitute a new spark plug.*
(b)   *Check the suppressor cap and HT lead for continuity.*
(c)   *Check for continuity to and from the ignition switch.*

(d)   *Check the ignition source coil.*
(e)   *Check the ignition pick-up coil.*
(f)   *Check the HT coil.*

3   If a check of the above items fails to cure the fault, the CDI must be assumed defective and renewed. Check first, however, that the unit's connector is not corroded or damaged and that the wiring to and from the unit is intact. Refer to the wiring diagram at the end of the Manual and make continuity checks through the ignition circuit to check for faulty wiring and connectors.

Ignition source coil – testing

4   Testing details are the same as described in Chapter 3, Section 6 for the earlier models, except for the location of the wiring block connector. This is the 5-pin connector on the left-hand side of the battery compartment (see photo 10.1); remove the frame centre cover to gain access. Make the test connections on the generator side of the connector as described in Chapter 3, noting the figure given in the specifications of this Chapter.

Ignition pickup coil – testing

5   Testing details are the same as described in Chapter 3, Section 7 for the earlier models, except for the location of the wiring block connector. This is the 5-pin connector on the left-hand side of the battery compartment (see photo 10.1); remove the frame centre cover to gain access. Make the test connections on the generator side of the connector as described in Chapter 3.

Ignition HT coil – location

6   The HT coil is mounted on the right-hand rear frame rail, just forward of the oil tank. Remove the under-seat storage compartment for access as described in Section 6. The coil can be removed by detaching the suppressor cap from the spark plug, disconnecting the wiring from the rear of the coil and removing the bolt which secures it to the frame rail.
7   Testing can be carried out as described in Chapter 3, Section 9.

---

6   Body panels: removal and refitting

Note: *The body panels are held together by a multitude of plastic tabs and clips which will break if overstressed. Pay particular attention to the removal sequence given because the removal order is critical in some cases. If a panel proves difficult to remove, make absolutely sure that all clips have been released before applying too much pressure. When refitting the panels, do not overtighten the retaining screws/bolts – such action may cause the panel to fracture around the mounting hole.*

Rear carrier

1   Raise the seat and remove the four nuts securing the carrier stays to the machine. Lift the carrier free.

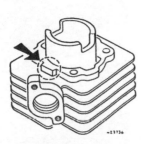

Fig. 7.2 Piston and cylinder barrel identification location (Sec 4)

5.1 CDI unit location

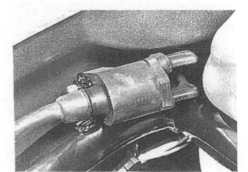

5.6 Ignition HT coil location

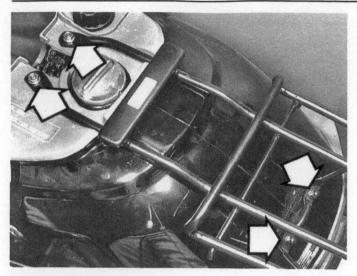

6.1 Rear carrier securing nuts (arrowed)

2   Refitting is a reversal of the removal operation. Note that the two domed nuts should be at the rear of the carrier and the plain nuts at the front.

Seat
3   Release the seat via the lock on the left-hand rear body panel. The seat hinge is secured by two nuts at the front of the storage compartment .

Under-seat storage compartment
4   Remove the rear carrier, seat and oil tank filler cap. Place a piece of plastic food wrap over the tank filler orifice to prevent the entry of dirt whilst the cap is removed.
5   Release the large slotted head fastener from the front upper edge (near the seat hinge location) and remove the two bolts at the base of the storage compartment. Lift the compartment out of the machine.
6   Refitting is a reverse of the removal procedure.

Frame centre cover
7   First remove the rear carrier, seat and the large slotted fastener from the top of the under-seat storage compartment. The centre cover is secured by a series of clips to the rear body panels and floorboard. Pull the cover free of the clips adjoining it to the body panels, and remove it forwards.
8   Refitting is a reversal of the removal procedure.

Frame lower covers
9   Remove the centre cover. Release the plastic screw-type fastener from the rear of the section and remove it together with its expanding insert.
10   Remove the two screws from the front edge of the section, pull this section outwards at its front edge and disengage the clip from the fillet in the middle section. Pull the section forwards to release it from the hooked clips on its rear section.
11   Refitting is a reversal of the removal procedure.

Rear body panels
12   The large right and left-hand panels must be removed together as an assembly. Start by removing the rear carrier, seat, centre cover and both frame lower covers. Remove the single bolt on each front lower corner and manoeuvre the complete assembly over the rear of the machine, rear light cluster and seat lock. Take care not to spread the front ends of the panel any more than is necessary for it to pass over the frame.
13   Refitting is a reverse of the removal procedure.

Floorboards
14   First remove the centre cover and frame lower covers as described above.
15   Remove the battery as described in Section 7. Remove the single bolt from the base of the battery compartment.
16   Remove the four main bolts visible from the upper face of the floorboards and prise the rear section of the floorboard/battery compartment free of the frame, being careful not to strain the wiring. Pull the floorboard rearwards and disengage it from the tags joining it to the legshield.

6.3 Remove two nuts to release seat and hinge

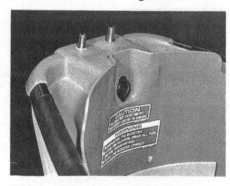

6.5 Large slotted fastener location in storage compartment

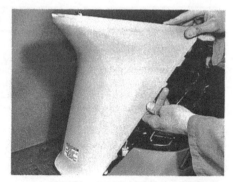

6.7 Frame centre cover is clipped in place

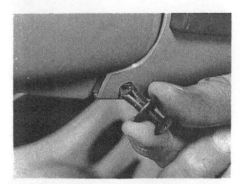

6.9 Remove special fastener at rear of frame lower covers ...

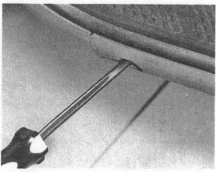

6.10a ... and two screws at front edge to release

6.10b Disengage clips on rear upper edge as lower cover is pulled forwards and withdrawn

6.12 Rear body panel front mounting bolt

6.15 Floorboard rear section retaining bolt can be accessed once battery has been removed

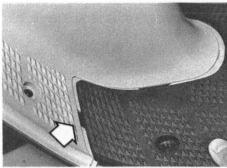

6.16 Floorboard front locating tag (arrowed)

6.17a Front cover is secured by three domed nuts at front ...

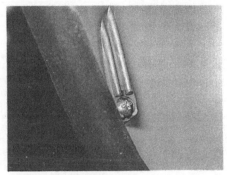

6.17b ... and two screws located under the front edge of the mudguard

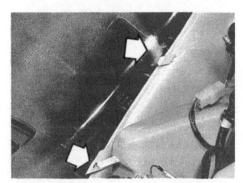

6.18 Align locating clips when refitting the front cover

6.20a Remove bolt in front storage compartment ...

6.20b ... two screws from lower section (near floorboard join) ...

6.20c ... and two screws at mudguard top corner to release legshield

Front cover

17   Remove the three domed nuts from the front of the cover and the two self-tapping screws from beneath the front edge of the mudguard. Press the front cover inwards at the rear join with the mudguard on each side and also at the joining edge with the legshield; this will serve to release the clips. Pull the cover off at the bottom edge first and then remove from the machine.

18   Refitting is a reversal of the removal procedure, but take care to align the clips as the cover is positioned.

Legshield

19   Remove the front cover, both frame lower covers and the floorboard.

20   Open the front storage compartment and remove the single bolt from the back of the legshield. Remove the two self-tapping screws from the lower edge of the legshield, and the two screws joining the legshield to the front mudguard. Pull the legshield rearwards carefully, so not to strain any wiring and spread the top part so that it will pass over the steering stem.

21   Refit in a reversal of the removal procedure, noting the locating ridge between the legshield and mudguard.

Front mudguard

22   First remove the front cover, both frame lower covers and the legshield as described above.

23   Remove the single nut retaining the mudguard to the electrical components bracket. Release the two nuts retaining the electrical components bracket to the steering stem and tie the bracket up out of the way.

24   Disconnect the front brake cable from its location on the wheel and pull the cable through its aperture in the mudguard.

25   Note that mudguard removal is possible with the wheel in place, but a certain amount of manoeuvring will be necessary to allow it to pass over the fork yokes. To permit this, the mudguard has a fillet at the back which when unclipped and hinged outwards will allow the rear ends to be spread outwards and passed over the front end of the machine. Position the rear bottom ends of the mudguard over the frame supports for the floorboard and raise the main section so that it passes

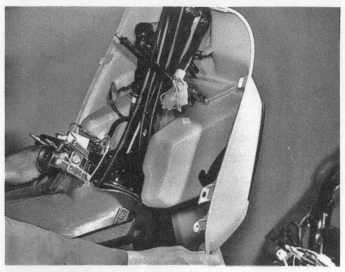

6.20d Legshield must be spread at top to allow it to pass over the steering column

over the fork yoke (see the accompanying photographs for details).

26   Refitting is a reversal of the removal procedure, noting that the fillet must be correctly positioned around the steering stem at the top and that it must be securely clipped into place. Pass the front brake cable through the mudguard, reconnect it and adjust it as described in Routine maintenance at the beginning of this Manual.

Handlebar nacelle

27   The handlebar nacelle may be removed independently of all other body panels.

28   Remove the rear view mirrors. The nacelle rear section is secured by three screws and three clips on its top edge (see photo 6.28a for location). Remove the screws and prise the clips free to release the rear cover. Unscrew the speedometer cable and disconnect the wiring block connectors to fully release.

29   To remove the front half of the nacelle, first remove the rear half, then release the headlamp bulbholder from the back of the reflector and similarly release both turn signal bulbholders from their locations. Remove the single screw at its bottom edge and the two screws which attach it to the handlebar brackets and withdraw the nacelle front half.

7   Battery: removal and refitting

1   The battery is of the sealed type, like the previous Visions, and is located behind the centre cover at the rear of the floorboard. Refer to Section 6 and remove the seat and frame centre cover for access.

2   To remove, disconnect its terminal connections (negative lead first) and remove the single screw securing its strap. The battery can now be manoeuvred from position, not forgetting to disconnect its fuseholder from the retaining clip. Refit in the reverse of the removal procedure, noting that the negative lead should be connected last and that the protective cover should fit correctly over the positive terminal.

3   Battery examination and charging can be carried out as described in Chapter 6, Section 4, but note the charge rates given in the Specifications of this Chapter.

6.23 Release mudguard nut (A) and electrical components bracket nuts (B)

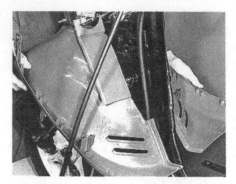

6.25a Release and fold back fillet at back of mudguard to allow it to be removed with the wheel in place

6.25b Mudguard removal requires the removal of the electrical components bracket and a little ingenuity to pass it over the steering stem

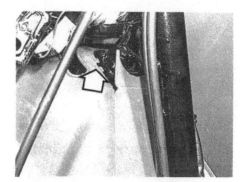

6.26 When refitting the mudguard, take care to snap the fillet correctly into place, particularly the top section – this must engage the clips shown when finally positioned

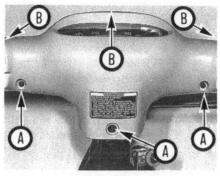

6.28a Nacelle rear section is held by three screws (A) and clips (B) at the points indicated

6.28b Showing the rear section retaining clip at the outer edge

7.2 Battery retaining strap screw

8.1 Regulator/rectifier unit location on electrical components plate

8   Regulator/rectifier unit: location and testing

1   The regulator/rectifier unit is located on a separate bracket extending from the steering stem. Access can be gained after removing the front cover as described in Section 6.
2   The unit is secured by a single bolt and has a plug-in type wiring connector. Testing can be carried out as described in Chapter 6, Section 6, referring also to the accompanying test table.

+ −	A	B	C	D
A		∞	3 – 50K	∞
B	∞			5-100K
C	∞	∝		∞
D	∞	5-100K	∞	

H.16832

Fig. 7.3 Regulator/rectifier unit test table (Sec 8)

9   Flywheel generator coils: testing

1   Testing details are exactly as specified in Chapter 3, Section 7, with the exception of the block connector location.
2   Remove the frame centre cover as described in Section 6 and identify the 5-pin connector on the left-hand side of the battery (see photo 10.1 for details). Disconnect the connector and make the resistance checks specified in Chapter 3 on the generator side of the connector.

10   Starter relay: location

1   The starter relay is mounted on top of the battery, next to the CDI unit. Remove the seat and centre cover to gain access. It is held in a plastic clip and can be removed by pulling it free once its wiring connector plug has been released.
2   Testing of the relay can be carried out as described in Chapter 6, Section 11.

10.1 Starter relay location (A). Note block connector location (B) for generator coils test

11   Resistors: location and testing

1   Two resistors are fitted, one from the lighting switch (identified by its pink wire) and the other from the automatic choke (identified by its green wire with black tube). Both resistors are bolted to the same mounting bracket as the regulator/rectifier unit. Remove the front cover as described in Section 6, for access.
2   If frequent headlamp bulb failure or problems with the automatic choke occur, check the appropriate resistor's resistance. Disconnect the wire connector to the resister and connect a multimeter set to the resistance (ohms) scale between the wire connector half on the resistor

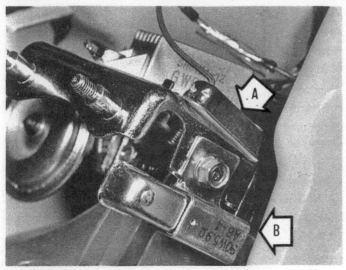

11.1 Resistors – automatic choke (A) and lighting (B)

side and earth. In the case of the lighting resistor the value should be 5.9 ohms and the automatic choke resistor should show a reading of 5.0 ohms.

12   Fuel gauge sender unit: removal, testing and refitting

1   Remove the rear body panels as described in Section 6.
2   Trace the three wires from the sender unit and disconnect them at the bullet connectors. Release the sender unit retaining ring by turning it with a pair of slip-jointed pliers or similar until its cut-outs align with the tabs on the tank surface. Withdraw the sender unit from the tank, taking care not to bend the float arm.
3   Using a multimeter set to the resistance (ohms) range take the following readings between the pairs of wires listed with the float in the full and empty positions.

Meter connections	Full position	Empty position
Yellow/white to green wire	33 ohms	533 ohms
Blue/white to green wire	533 ohms	33 ohms
Blue/white to yellow/white wire	600 ohms	600 ohms

4   If the meter readings are outside of the expected results, or the

meter reading fluctuates wildly as the float is raised/lowered, this may be due to corrosion of the resistance windings or float arm. Nothing is lost by attempting to clean up corroded windings as described in Chapter 6, Section 15, but if this fails to cure the fault, the sender unit must be renewed.
5   Check the condition of the sealing ring before inserting the sender unit in the tank, and renew it if necessary. Align the tank tabs with the sender unit cut-outs to fully insert it, and then twist the sender away from the tabs to secure it (the correct position is when the arrows on the sender and tank are in alignment – see photo 15.1a in Chapter 6).

13   Headlamp: removal and refitting

1   To gain access to the bulb, first remove the nacelle rear section as described in Section 6. Note that there is no need to disconnect the wiring and speedometer cable for bulb access. Twist the bulbholder anticlockwise to release it from the retainers and withdraw the bulbholder complete with headlamp bulb from the reflector. The bulb is a bayonet fitting in its holder; twist anticlockwise to remove.
2   If the headlamp reflector unit requires removal, note that it is secured to the nacelle front half by four screws. Remove the nacelle to gain access.
3   Two beam adjuster screws are located in the front half of the nacelle, just below the headlamp.

14   Turn signals: general

Front turn signal bulbs – renewal
1   To gain access to either turn signal bulb, remove the nacelle rear half as described in Section 6. Twist the bulbholder to remove it from the reflector unit. The bulb is a bayonet fitting in its holder; twist anticlockwise to remove.
2   To remove the reflector unit, remove the nacelle rear half as described in Section 6, and then remove the two screws which retain the unit to the nacelle front half. Disconnect the bulbholder and remove through the nacelle aperture. Remove the single screw to separate the lens from the reflector .

Turn signal relay – location
3   The relay is located in the instrument panel. Remove the nacelle rear half to gain access.

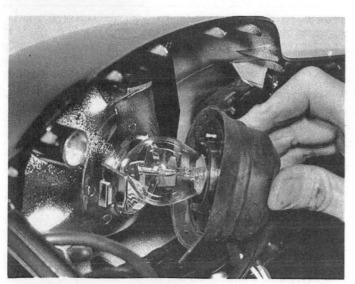

13.1a Headlamp bulbholder removal

13.1b Headlamp bulb removal from holder

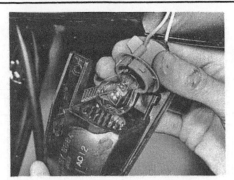

14.2a Remove two screws to release front turn signal reflector units

14.2b Turn signal bulbholder removal from reflector

14.3 Turn signal relay location in instrument panel

15   Instrument panel: removal and refitting

1   To gain access to the instruments, warning lights or switches, first remove the nacelle rear half as described in Section 6. Unscrew the speedometer cable and disconnect the wiring block connectors to allow the instruments and nacelle rear half to be moved to the workbench.

2   The four warning lamp/illumination bulbholders are a push fit in the back of the instruments.

3   To dismantle the instrument assembly, first disconnect the wiring from the back of the fuel gauge (having noted the wire positions) and the block connector from the turn signal relay. Remove the three screws retaining the instruments to the nacelle and separate the two components. Remove the five screws from the top face and lift off the instrument cover and sealing ring. Remove the four screws and release the front face/speedometer unit from the instrument housing. Access can now be gained to the oil level LED and fuel gauge.

4   Reassemble in a reverse of the removal sequence, taking particular care over the positioning of the two dampers between the front face and cover, and the cover sealing ring.

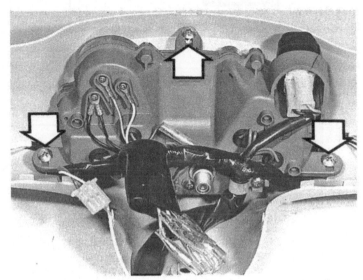

15.3 Rear view of instruments showing three instrument-to-nacelle screws

16   Handlebar switches: removal and refitting

1   Access to all handlebar switches can be gained after removing the nacelle rear half as described in Section 6. Where a detachable type connector is fitted, pull the wiring free from the rear of the switch. Each switch is held in the nacelle by plastic ears on the side of the switch body, which when depressed will allow the switch to be removed from the front of the nacelle.

2   When refitting the switches, make sure that they snap correctly back into position in the nacelle.

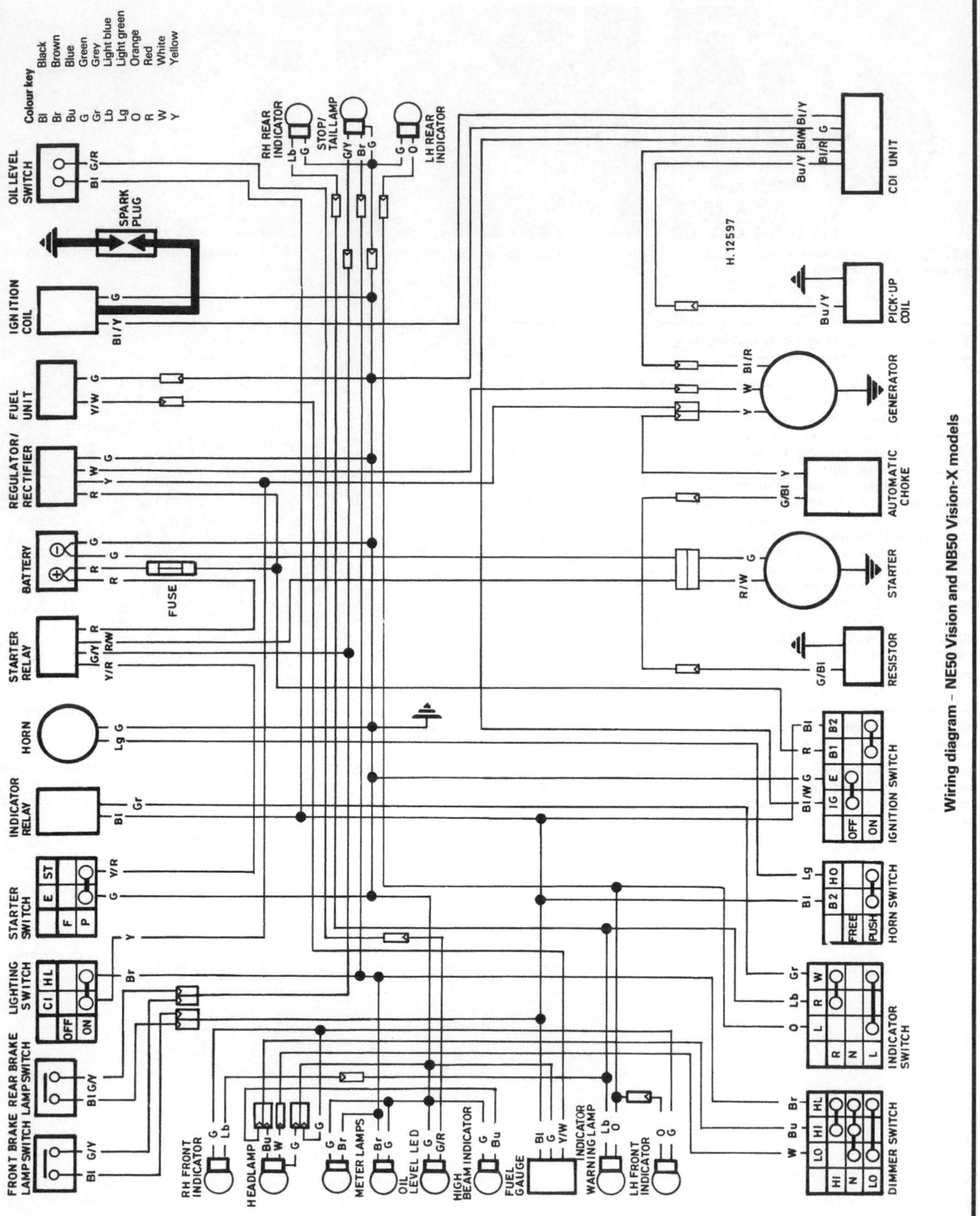

Colour key

Bl	Black
Br	Brown
Bu	Blue
G	Green
Gr	Grey
Lb	Light blue
Lg	Light green
O	Orange
R	Red
W	White
Y	Yellow

OIL LEVEL SWITCH

RH REAR INDICATOR

STOP/ TAIL LAMP

LH REAR INDICATOR

CDI UNIT

SPARK PLUG

IGNITION COIL

PICK-UP COIL

H.12597

FUEL UNIT

REGULATOR/ RECTIFIER

GENERATOR

AUTOMATIC CHOKE

BATTERY

FUSE

STARTER

STARTER RELAY

RESISTOR

HORN

IGNITION SWITCH

INDICATOR RELAY

HORN SWITCH

STARTER SWITCH

LIGHTING SWITCH

INDICATOR SWITCH

FRONT BRAKE LAMP SWITCH

REAR BRAKE LAMP SWITCH

RH FRONT INDICATOR

HEADLAMP

METER LAMPS

OIL LEVEL LED

HIGH BEAM INDICATOR

FUEL GAUGE

WARNING LAMP

LH FRONT INDICATOR

DIMMER SWITCH

Wiring diagram – NE50 Vision and NB50 Vision-X models

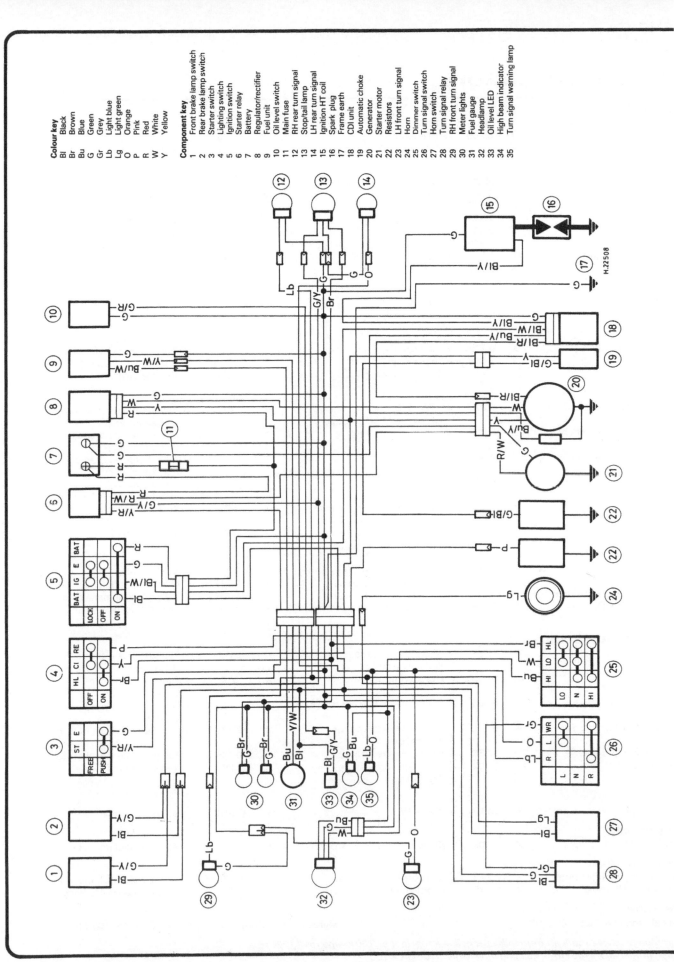

Colour key

Bl	Black
Br	Brown
Bu	Blue
G	Green
Gr	Grey
Lb	Light blue
Lg	Light green
O	Orange
P	Pink
R	Red
W	White
Y	Yellow

Component key

1. Front brake lamp switch
2. Rear brake lamp switch
3. Starter switch
4. Lighting switch
5. Ignition switch
6. Starter relay
7. Battery
8. Regulator/rectifier
9. Fuel unit
10. Oil level switch
11. Main fuse
12. RH rear turn signal
13. Stop/tail lamp
14. LH rear turn signal
15. Ignition HT coil
16. Spark plug
17. Frame earth
18. CDI unit
19. Automatic choke
20. Generator
21. Starter motor
22. Resistors
23. LH front turn signal
24. Horn
25. Dimmer switch
26. Turn signal switch
27. Horn switch
28. RH front turn signal
29. Meter lights
30. Fuel gauge
31. Headlamp
32. Oil level LED
33. High beam indicator
34. Turn signal warning lamp

H.22508

Wiring diagram – SA50 Vision Met-in model

Conversion factors

Length (distance)

	X			X		
Inches (in)	X	25.4	= Millimetres (mm)	X	0.0394	= Inches (in)
Feet (ft)	X	0.305	= Metres (m)	X	3.281	= Feet (ft)
Miles	X	1.609	= Kilometres (km)	X	0.621	= Miles

Volume (capacity)

	X			X		
Cubic inches (cu in; in³)	X	16.387	= Cubic centimetres (cc; cm³)	X	0.061	= Cubic inches (cu in; in³)
Imperial pints (Imp pt)	X	0.568	= Litres (l)	X	1.76	= Imperial pints (Imp pt)
Imperial quarts (Imp qt)	X	1.137	= Litres (l)	X	0.88	= Imperial quarts (Imp qt)
Imperial quarts (Imp qt)	X	1.201	= US quarts (US qt)	X	0.833	= Imperial quarts (Imp qt)
US quarts (US qt)	X	0.946	= Litres (l)	X	1.057	= US quarts (US qt)
Imperial gallons (Imp gal)	X	4.546	= Litres (l)	X	0.22	= Imperial gallons (Imp gal)
Imperial gallons (Imp gal)	X	1.201	= US gallons (US gal)	X	0.833	= Imperial gallons (Imp gal)
US gallons (US gal)	X	3.785	= Litres (l)	X	0.264	= US gallons (US gal)

Mass (weight)

	X			X		
Ounces (oz)	X	28.35	= Grams (g)	X	0.035	= Ounces (oz)
Pounds (lb)	X	0.454	= Kilograms (kg)	X	2.205	= Pounds (lb)

Force

	X			X		
Ounces-force (ozf; oz)	X	0.278	= Newtons (N)	X	3.6	= Ounces-force (ozf; oz)
Pounds-force (lbf; lb)	X	4.448	= Newtons (N)	X	0.225	= Pounds-force (lbf; lb)
Newtons (N)	X	0.1	= Kilograms-force (kgf; kg)	X	9.81	= Newtons (N)

Pressure

	X			X		
Pounds-force per square inch (psi; lbf/in²; lb/in²)	X	0.070	= Kilograms-force per square centimetre (kgf/cm²; kg/cm²)	X	14.223	= Pounds-force per square inch (psi; lbf/in²; lb/in²)
Pounds-force per square inch (psi; lbf/in²; lb/in²)	X	0.068	= Atmospheres (atm)	X	14.696	= Pounds-force per square inch (psi; lbf/in²; lb/in²)
Pounds-force per square inch (psi; lbf/in²; lb/in²)	X	0.069	= Bars	X	14.5	= Pounds-force per square inch (psi; lbf/in²; lb/in²)
Pounds-force per square inch (psi; lbf/in²; lb/in²)	X	6.895	= Kilopascals (kPa)	X	0.145	= Pounds-force per square inch (psi; lbf/in²; lb/in²)
Kilopascals (kPa)	X	0.01	= Kilograms-force per square centimetre (kgf/cm²; kg/cm²)	X	98.1	= Kilopascals (kPa)
Millibar (mbar)	X	100	= Pascals (Pa)	X	0.01	= Millibar (mbar)
Millibar (mbar)	X	0.0145	= Pounds-force per square inch (psi; lbf/in²; lb/in²)	X	68.947	= Millibar (mbar)
Millibar (mbar)	X	0.75	= Millimetres of mercury (mmHg)	X	1.333	= Millibar (mbar)
Millibar (mbar)	X	0.401	= Inches of water (inH₂O)	X	2.491	= Millibar (mbar)
Millimetres of mercury (mmHg)	X	0.535	= Inches of water (inH₂O)	X	1.868	= Millimetres of mercury (mmHg)
Inches of water (inH₂O)	X	0.036	= Pounds-force per square inch (psi; lbf/in²; lb/in²)	X	27.68	= Inches of water (inH₂O)

Torque (moment of force)

	X			X		
Pounds-force inches (lbf in; lb in)	X	1.152	= Kilograms-force centimetre (kgf cm; kg cm)	X	0.868	= Pounds-force inches (lbf in; lb in)
Pounds-force inches (lbf in; lb in)	X	0.113	= Newton metres (Nm)	X	8.85	= Pounds-force inches (lbf in; lb in)
Pounds-force inches (lbf in; lb in)	X	0.083	= Pounds-force feet (lbf ft; lb ft)	X	12	= Pounds-force inches (lbf in; lb in)
Pounds-force feet (lbf ft; lb ft)	X	0.138	= Kilograms-force metres (kgf m; kg m)	X	7.233	= Pounds-force feet (lbf ft; lb ft)
Pounds-force feet (lbf ft; lb ft)	X	1.356	= Newton metres (Nm)	X	0.738	= Pounds-force feet (lbf ft; lb ft)
Newton metres (Nm)	X	0.102	= Kilograms-force metres (kgf m; kg m)	X	9.804	= Newton metres (Nm)

Power

	X			X		
Horsepower (hp)	X	745.7	= Watts (W)	X	0.0013	= Horsepower (hp)

Velocity (speed)

	X			X		
Miles per hour (miles/hr; mph)	X	1.609	= Kilometres per hour (km/hr; kph)	X	0.621	= Miles per hour (miles/hr; mph)

Fuel consumption

	X			X		
Miles per gallon, Imperial (mpg)	X	0.354	= Kilometres per litre (km/l)	X	2.825	= Miles per gallon, Imperial (mpg)
Miles per gallon, US (mpg)	X	0.425	= Kilometres per litre (km/l)	X	2.352	= Miles per gallon, US (mpg)

Temperature

Degrees Fahrenheit = (°C x 1.8) + 32   Degrees Celsius (Degrees Centigrade; °C) = (°F – 32) x 0.56

It is common practice to convert from miles per gallon (mpg) to litres/100 kilometres (l/100km), where mpg (Imperial) x l/100 km = 282 and mpg (US) x l/100 km = 235

Index

A

Adjustment:-
  brakes 21, 94
  carburettor 65
  idle speed 20
  spark plug 22, 116
  steering head bearings 76
Air filter 20, 65
Automatic choke 61

B

Battery 101, 120
Bearings:-
  engine 35
  front wheel 92
  rear wheel 94
  steering head 23, 76
Bleeding the engine lubrication system 21
Body panels 19, 83, 117 – 120
Brakes:-
  adjustment 21, 94
  examination and renovation 94
  fault diagnosis 16
  specifications 90, 115
  stop lamp 110
  stop lamp switch 105
  wear check 94

C

Cables 23
Carburettor:-
  adjustment 20, 65
  cold start device (automatic choke) 61
  examination and renovation 62
  refitting 53, 62
  removal 30, 62
CDI unit 73, 117
Charging system 102, 121
Clutch – centrifugal:-
  examination and renovation 40
  refitting 47
  removal 32
  specifications 26
Cold start device (automatic choke) 61
Conversion factors 126
Crankcases:-
  joining 45
  separating 33
Crankshaft 34
Cylinder barrel and head:-
  examination and renovation 35, 117
  refitting 50
  removal 31

D

Decarbonisation:-
  engine 22
  exhaust 22, 66
Dimensions – model 5, 114
Drive pulley – transmission:-
  belt 32, 40 ,47
  drive pulley 33, 37, 47
  driven pulley 32, 40, 47

E

Electrical system:-
  battery 101, 120
  charging system 102, 121
  checks 100
  fault diagnosis 16, 17
  fuel gauge and sensor 106, 122
  fuse 102
  headlamp 101, 122
  horn 110
  instrument panel 108, 123
  oil level snesor 106
  regulator/rectifier 102, 121
  resistor(s) 102, 121
  specifications 99
  starter system 102, 104
  switches 8, 105, 123
  turn signal lamps and relay 110, 117
  wiring diagrams 124, 125
Engine:-
  bearings 35
  crankcases 33, 45
  crankshaft 34
  cylinder barrel and head 31, 35, 50, 117
  decarbonisation 22
  dismantling – general 27
  examination and renovation – general 34
  fault diagnosis 12
  flywheel 28, 52
  mountings – check 55
  oil level check 18
  oil pump 30, 52, 67
  oil seals 35
  piston 31, 36, 50, 117
  piston rings 36
  reassembly – general 45
  refitting into the frame 55
  removal from the frame 27
  specifications 25, 114
  starting and running 57
Exhaust system:-
  decarbonisation 22, 66
  refinishing 66
  refitting 53, 66
  removal 27, 66

F

Fault diagnosis:-
  brakes 16
  clutch 14
  electrical 16-17
  engine 12
  frame 15-16
  fuel system 13
  lubrication 14
  transmission 15
Floorboard 19, 83, 118
Flywheel generator:-
  refitting 52
  removal 28
  testing 102, 121

Frame 82
Front brake:-
  adjustment 21, 94
  examination and renovation 94
  fault diagnosis 16
  specifications 90, 115
  stop lamp 110
  stop lamp switch 105
  wear check 94
Front cover 19, 83, 119
Front suspension:-
  check 22
  examination and renovation 79
  removal and refitting 76
  specifications 75, 115
Front wheel:-
  bearings 92
  check 21
  examination and renovation 90
  removal and refitting 91
Fuel:-
  filter 20
  gauge and sensor 106, 122
  pipes 21, 61
  tank 60
  tap 60
Fuel system:-
  carburettor 20, 30, 53, 61, 62
  fault diagnosis 13
  reed valve 30, 52, 66
  specifications 59, 115
Fuse 102

G

Gearbox see Transmission
Generator – flywheel:-
  refitting 52
  removal 28
  testing 102, 121

H

Handlebar:-
  nacelle 83, 120
  switches 105, 123
Headlamp 109, 122
Horn 110
HT coil and lead 74, 117

I

Ignition system:-
  CDI unit 73, 117
  checks 70, 72, 117
  HT coil and lead 74, 117
  pickup coil 72, 117
  source coil 72, 117
  spark plug 23, 70, 71, 116
  specifications 68, 115
  switch 88, 105
  timing 74
Instrument panel 108, 123

K

Kickstart:-
examination and renovation 45
lever 89
refitting 49
removal 32

L

Lamps 99
Legshield 19, 83, 119
Lubrication:-
control cables 23
engine:
bleeding 67
check 18, 21
pump 30, 52, 67
fault diagnosis 14
specifications 18
transmission 55

M

Main bearings 35
Maintenance – routine 18 – 24, 116

O

Oil – engine:-
filter 67
level:
check 18
sensor 106
lines 21, 67
pump:
refitting 52, 67
removal 30, 67
system bleeding 67
tank 66
Oil – transmission 55
Oil seals 35

P

Pickup coil 72, 117
Piston:-
examination and renovation 36, 117
refitting 50
removal 31
rings 36
Pulley – transmission:-
drive 33, 37, 47
driven 32, 40, 47

R

Rear brake:-
adjustment 21, 94
examination and renovation 94
fault diagnosis 16
specifications 90, 115
stop lamp 110
stop lamp switch 105
wear check 94
Rear suspension:-
check 22
pivots 80
unit 81

Rear wheel:-
bearings 94
examination and renovation 94
refitting 53, 94
removal 28, 94
Reed valve:-
examination and renovation 65
refitting 52
removal 30
Regulator/rectifier 102, 121
Resistor(s) 102, 121
Rings – piston 36
Routine maintenance 18 – 24, 116

S

Safety first! 7
Seat 88, 118
Side panels 19, 83, 118
Source coil – ignition 72, 117
Spark plug:-
check 72, 116
colour condition chart 71
renewal 23, 70
Specifications:-
brakes 90, 115
electrical system 99, 115
engine 25, 114
frame and suspension 75, 115
fuel system 59, 115
ignition system 68, 115
lubrication system 18, 59
routine maintenance 18
transmission 26, 114
tyres 90, 115
wheels 90
Speedometer:-
cable 89
drive 89, 92
head 89, 108, 123
Stand 88
Starter motor:-
fault diagnosis 17
refitting 52
relay 104, 121
removal 30
system 102
testing 104
Steering:-
check 19
head:
bearings 23
removal and refitting 76
lock 88
Stop lamp 110
Suspension:-
check 22
fault diagnosis 16
front 76, 79
rear 80, 81
specifications 75, 115
Switches:-
brake stop lamp 105
handlebar 105, 123
ignition 88, 105

T

Tail lamp 110
Timing – ignition 74
Tools 8
Torque wrench
settings 10, 26, 75, 90, 114, 115
Transmission:-
centrifugal clutch:
examination and renovation 40
refitting 47
removal 32
checks 23
cover:
refitting 49
removal 32
drive belt:
examination and renovation 40
refitting 47
removal 32
drive plulley:
examination and renovation 37
refitting 47
removal 33
driven pulley:
examination and renovation 40
refitting 47
removal 32
final reduction gearbox:
examination and renovation 40
refitting 46
removal 33
oil 18, 55
specifications 26, 114
Turn signal lamps and relay 110, 122
Tyres:-
pressures 19, 90
removal and refitting 96
size 90, 115
valves 98

V

Valve – reed:
examination and renovaion 65
refitting 52
removal 30
Valve – tyre 98
Voltage regulator 102, 121

W

Weights 5, 114
Wheels:-
check 21
front:
bearings 92
examination and renovation 90
removal and refitting 91
rear:
bearings 94
examination and renovation 94
refitting 53, 94
removal 28, 94
Wiring diagrams 124, 125